C0-ANY-193

# COMMON SENSE MANUFACTURING

## BECOMING A TOP VALUE COMPETITOR

The BUSINESS ONE IRWIN/APICS Series
in Production Management

Supported by the American Production
and Inventory Control Society

OTHER BOOKS PUBLISHED IN THE BUSINESS ONE IRWIN/APICS
SERIES IN PRODUCTION MANAGEMENT

Attaining Manufacturing Excellence *Robert W. Hall*

Bills of Materials *Hal Mather*

Production Activity Control *Steven A. Melnyk and Phillip L. Carter*

Manufacturing Planning and Control Systems, Second Edition
*Thomas E. Vollmann, William Lee Berry, D. Clay Whybark*

The Spirit of Manufacturing Excellence *Ernest Huge*

Strategic Manufacturing: Dynamic New Directions for the 1990's *Patricia E. Moody*

Total Quality: An Executive's Guide for the 1990s
*The Ernst & Young Quality Improvement Consulting Group*

Forecasting Systems for Operations Management
*Stephen A. DeLurgio and Carl D. Bahme*

# COMMON SENSE MANUFACTURING

## BECOMING A TOP VALUE COMPETITOR

*James A. Gardner, CPIM*

*The Business One Irwin/APICS Series in Production Management*

**BUSINESS ONE IRWIN**
**Homewood, Illinois 60430**

This publication is designed to provide accurate and authoritative information in regard to the subject matter covered. It is sold with the understanding that neither the author nor the publisher is engaged in rendering legal, accounting, or other professional service. If legal advice or other expert assistance is required, the services of a competent professional person should be sought.

*From a Declaration of Principles jointly adopted by a Committee of the American Bar Association and a Committee of Publishers.*

Senior editor: Jeffrey A. Krames
Project editor: Jean Roberts
Production manager: Ann Cassady
Compositor: Carlisle Communications, Ltd.
Typeface: 11/13 Palatino
Printer: R. R. Donnelley & Sons Company

**Library of Congress Cataloging-in-Publication Data**

Gardner, James A. (James Allen)
Common sense manufacturing : becoming a top value competitor / James A. Gardner.
p. cm.
Includes index.
ISBN 1-55623-527-5
1. Production management. 2. Production control. 3. Materials management. I. Title.
TS155.G338 1991
658.5'1—dc20 91-14410

*Printed in the United States of America*

1 2 3 4 5 6 7 8 9 0 DOC 8 7 6 5 4 3 2 1

# PREFACE

Modern America is totally caught up in buzzword technology. The use of jargon that began not long ago with governmental bodies and agencies has become interwoven into every fabric of today's society. Clever catch words, phrases, and acronyms are evident in every walk of life. Manufacturing is no different. Production and Inventory Control have generated their own special buzzwords. The most popular ones and those that are used universally include Manufacturing Resource Planning and/or MRP II, JIT, and CIM.

The purpose of jargon, as everyone knows, is to prove that the individual using it is well versed in its relevancy and application. As everyone further knows, however, this is not a fact. The objective of this book is to look beyond the buzzwords into the origin, reason, and intent for their existence. The idea herein is to strip away much of the complexity surrounding the various concepts and techniques and explain them in everyday plain language. It is most important to know when to use buzzwords, not merely how. Simply mouthing the magic words is not enough. This, in its most basic and unadulterated form, is what is meant by *Common Sense Manufacturing*.

Whenever one reads an article from some current periodical concerning manufacturing, operations, or material, he or she can be almost certain of finding two things. First, the article will contain a liberal sprinkling of some of the more common and currently prevalent buzzwords; and second, the author, in all likelihood, will be from IBM or Xerox or Steelcase or Eastman Kodak, or some other company whose name is a household word and one of the firms included in *Fortune's* Top 50 Manufacturing Companies. Of course, there is nothing wrong with this when created in the right frame of mind. Such an article should be accepted as a contribution to a continuing body of knowledge. As far as practical application goes, however, its value is often virtually nil. There is a mentality created over the last several generations that equates bigness with "bestness." To submit that a system or systems devised by a factory with 2,500 people, nearly unlimited resources, and an almost inexhaustible supply of talent to tap into can somehow be used by a factory having 250 employees and none of the above advantages is just not tenable.

Recalling an APICS chapter meeting I attended a year or so ago demonstrates this point. The speaker was from a nearby maker of office furniture and the presentation was interesting and well done. The topic was on-time deliveries and the individual who had engineered the system could be justly proud of such an accomplishment. Apparently, however, it never occurred to the group in attendance that the adaptability of this particular on-time delivery system to any other manufacturer, especially a smaller firm that had neither the same resources nor an equal volume of product to be moved, was highly improbable, if not impossible.

If this book were to be dedicated to anyone, it would be dedicated to the myriad thousands of smaller firms that go unrepresented in a world where they remain anonymous and largely ignored. The same goes for the ideas and concepts contained herein. They are designed to fit into situations where they can be supportive, not because someone else employed them successfully.

Rare is the case where an entire system should be abandoned in favor of a better one. Retaining effective portions and adding improvements as called for is certainly preferable to throwing out the "baby with the bath water." One reason for system implementation failure is the failure to realize the importance of retaining and improving . To attempt to force something new and different on an unreceptive audience is an exercise in futility.

It would be quite a remission not to give credit where credit is due. The idea as well as the term *Common Sense Manufacturing* is the brainchild of my former partner, John J. Green, of Green and Associates. In addition, as a matter of courtesy John wrote the Foreword, which describes the framework of ideas that comprise Common Sense Manufacturing as well as the reasons that sparked its development.

While I borrowed an idea or two to build a continuum, the complete text in its entirety, as well as the opinions expressed therein, are exclusively my own. If any sacred cows were gored in the process, it was probably intentional. As such, no apology is offered.

James A. Gardner

# FOREWORD

As 1992 approaches, general and operating managers are confronted on one hand by realities of intensified worldwide competition and on the other by an "alphabet soup" of buzzwords and acronyms, all hailed as *the* solution to competitiveness or productivity problems. Yet for each productive CIM (computer-integrated manufacturing) installation, there are many cases where CIM has fallen short—or flat out did *not* apply. Similarly, truly successful SPC (statistical process control) programs and MRP (manufacturing resource planning) installations are in the minority: They simply have not yielded the advertised benefits. There are many reasons for this shortfall. Most have nothing to do with the merits of each technology but rather with its misapplication or the failure to first build an adequate organizational foundation.

To achieving the competitive edge, I advocate *Common Sense Manufacturing*, an emphasis on "blocking and tackling" and built around realistic principles:

- Common sense manufacturing is customer-focused because competitiveness is ultimately judged by current and prospective customers.
- Common sense manufacturing is an integrated system: Quality alone, price alone, delivery alone are all inadaquate to meet the customer's demand for value.
- Common sense manufacturing is a *living* system, meaning it is ultimately people-dependent. Reliance on high-tech investment to solve productivity problems is as foolhardy as relying on a motivational solution alone. People must know *what* to do and *how* to do it. Developing the competence of these human resources is essential.
- Common sense manufacturing utilizes technologies appropriate to the company, its people, and the market it serves. We do not believe that money is an automatic solution to competitiveness and productivity concerns.
- Finally, customer-focused competitiveness is an unending journey which begins with flexible strategic plans.

**John J. Green**

# CONTENTS

# LIST OF FIGURES

# PART 1

# PERSPECTIVE

# CHAPTER 1

# THE CUSTOMER FOCUS

Current management theory has it—especially if one subscribes to the Peter Drucker school of thought—that a business exists for only one reason: to create a customer. Of course this implies—although it is not expressly stated—that the customer, once created, must be maintained. If we hold this to be a fundamental concept, then in order for an enterprise to sustain itself and provide for long-term growth, it must be customer-focused. There is no alternative to this precept, and commitment to it cannot be overemphasized. It is not merely the first priority, it is the *only* priority. All else is subordinate.

The primary method for focusing on the customer is to provide him or her with superior value. This can be accomplished through the management of five key areas:

- Quality: strict adherence to the highest industry standards.
- Delivery: the goal of 100 percent on-time delivery.
- Cost: competitive advantage gained through efficient operations.
- Service: technical support supplied by your own team of specialists.
- Product: a thorough knowledge of the product and processes including its functions.

When one picks up a current magazine article on manufacturing or operations, customer service, if it is mentioned at all, is assumed to be more or less a given—that is, having performed his or her duty, the author launches into a detailed disquisition on a particular system or set of techniques that, if implemented correctly and administered wisely, would work to the benefit of all, maybe even to the customer. The underlying idea is that if everything comes together as planned, improved customer service will result as a matter of course. This is the wrong approach—it is too company-focused. A better approach is always to keep the customer at the center of attention and concentrate on fulfilling his or her wants and needs. If, along the way, we can develop techniques that will aid us in our efforts, then by all means we should put them into use. This is what is meant by common sense manufacturing: It is a people-driven system.

The decade of the 1980s was not a particularly happy or prosperous time for much of American industry—especially the manufacturing sector. The decade began with the country mired in a nasty recession, followed by a period in which the dollar became extremely expensive in relation to the other major world currencies. It almost seemed like we had awakened to find ourselves in a crisis. The United States had simply lost its ability to compete. The guiding principles of the 1960s and 70s no longer applied in the 80s.

All of this turmoil gave a new dimension to the focus on the customer. We are competing in a market that is truly worldwide in scope. No longer are our concerns only with the national scene and international trade but with foreign transplants competing directly with us in a market we had always considered our own.

As the 1990s dawn, however, improvements are becoming increasingly evident. There is light at the end of the tunnel. American competency, ingenuity, and our pioneering attitude will lead us to regain our position of prominence. Of this, I am sure.

Finally, our traditional view of the customer will have to be expanded. Our customer is not only the ultimate user of our finished product but also the next operation in our internal factory being supplied by a preceding operation. The needs of quality, quantity, uniformity, and timeliness are exactly the same, regardless of user. Customer-related requirements, the bedrock upon which all business activity should be based, form the core of common sense manufacturing, as described in this book.

# PART 2

# EVOLUTION

# CHAPTER 2

# A LONG NIGHT'S JOURNEY INTO DAY

## HISTORICAL PERSPECTIVE

The state of production and inventory management today is certainly a far cry from what it was in its infancy. During its long, and mostly rocky, evolutionary process, a body of knowledge has finally been developed that is truly an effective way of managing today's factory. Obviously this was not always true, and even today too many companies are still subscribing to the elements of "scientific inventory management."

From a historical standpoint, the growth of this discipline presents some interesting anomalies. The boom years following World War II were remarkable and are unlikely ever to be equalled. An entire generation of pent-up demand brought on by the Depression, when there was no money to spend, and then by the war years, when there was nothing to spend it on, finally found release and lasted until well into the 1950s. As the economy returned to relative normalcy, manufacturing companies, in order to retain their competitive edge, began to examine their operations in an effort to determine how best to streamline them. At this point Production Control was mainly a clerical function, wherein the plant manager or floor supervisor told the scheduler what to schedule, and Inventory Control, because inventory cost money, was logically handled by the Accounting Department.

With meager principles and techniques to guide corporate managers, many—often conflicting—theories arose as to how best to harness this growing, largely inefficient monster known as manufacturing. This brought about the heyday of Operations Research. It was believed that solutions to complex business problems lay in the construction of intricate, and sometimes incomprehensible, mathematical equations. This ushered in a period many termed *Scientific Inventory Management* (SIM). Although these views are not widely held today, some practitioners still show a reluctance to discard them as outmoded. Old habits die hard!

What has happened during the intervening years has been to keep the good parts, like forecasting; modify other concepts, like the critical ratio; and discard the many techniques that proved unworkable in practice, second-order, or even third-order exponential smoothing being a case in point.

While all this was going on, what was taking place in the typical manufacturing facility? An order would be received by customer service, order entry, or whatever (or a forecasted quantity would come due); the order would then be typed up in a manufacturing order and sent out to Production Control. A copy would also be sent to Engineering for identifying the correct bill of material. Production Control, taking into consideration lead time, would schedule shop orders onto the floor and send requisitions for buy-outs to Purchasing. Figure 2–1 depicts the schematic flow.

The situation out on the shop floor was somewhat more chaotic. Because of schedule changes, material shortages, equipment downtime, and various other reasons, dates on shop orders no longer bore any resemblance to when they were needed—not to mention when they were being produced—and a team of expeditors was required to pull what was actually needed through the processes. And because what was being received from vendors was in many cases not what was actually needed, a "hot list" of necessary parts was generated for Purchasing to expedite. When the hot list became too long, a second list became the "hot, hot list"; and when that became too long, as it often did, it became the "real-scorcher list."

A similar condition existed in work-in-progress on the factory floor. All those component parts that were most critical received a red tag. When practically everything in the shop had a red tag, the positively crucial parts—the hot, hot list—were also affixed with a green tag. The final insanity occurred when

**FIGURE 2–1**
**Order Flow Schematic**

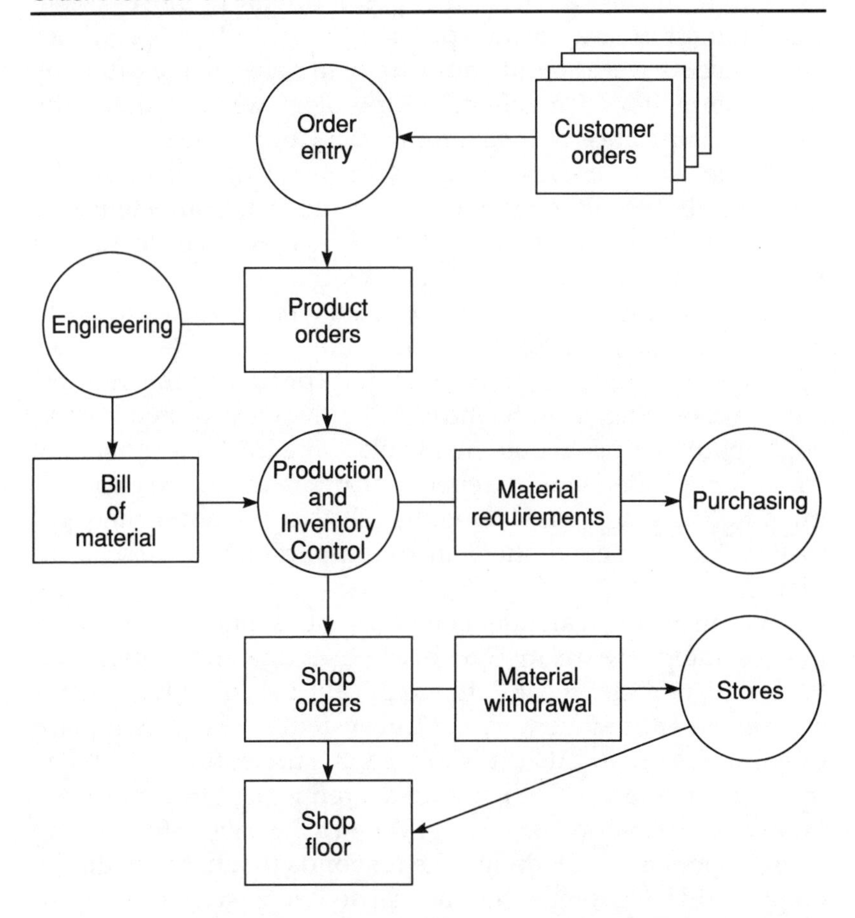


an expeditor had to tell his shift foreman, "I can't put a yellow tag on the red-tagged part because it doesn't have a green tag yet."

Needless to say, this led to burgeoning inventories, poor customer service, and everyone thinking everybody else was doing a lousy job of running things. In retrospect, the problem was obvious: The company was no longer running production; production was running the company. No one ever stopped to ask, "Are we asking the factory to produce more than it is capable of producing?" Things were much too hectic for that.

Company officials were in a quandary as to the best way of extricating themselves from this predicament. In most cases it was thought to be a "people problem." After all, the principles and concepts were sound and were being used successfully by their competitors. Therefore, the problem was obviously the lack of competence of their Production Control people.

If the truth were known, everyone was suffering under the same handicaps, although few would admit it. Competence, or its lack, had little to do with it. Still, it was easier to replace people than it was to replace systems.

Let's take a look at a more or less typical case. The president or CEO of a company attends a seminar on state-of-the-art methods of scheduling production and controlling inventories. After the seminar, with his head full of newly acquired knowledge about material requirements planning (MRP), exponential smoothing, independent-dependent demand, discrete-lot sizing, and the whole nine yards, he tells his personnel manager to find him someone adept in the application of these techniques.

When a likely candidate emerges, he is quite positive during his interview, relative to his knowledge and experience. When queried about MRP, the applicant describes how he instituted an entire inventory control system at his current place of employment and that it is extremely successful. In fact—he tells the interviewer—he has added a refinement to the concept known as part-period balancing. When asked about forecasting using exponential smoothing, he responds that not only did he employ this technique but he carried it a step further, to second-order smoothing to detect trends. He then describes how he controlled stock levels through the order point method, with safety stock amounts set using the mean absolute deviation. He concludes by expounding on critical path, critical ratio, and all the other popular buzzwords.

The effect of these new systems is easy to see when one looks forward 24–36 months after they are set in place. An order comes into the company through the usual channels, and a manufacturing order is issued and sent to Production Control, which then issues shop orders to the various work stations.

When the order is six weeks past due, the customer completely frustrated, and the company on its third production and inventory control manager, the expeditor finally forces the order through the shop—and the problems of poor customer service, top-heavy inventories, and completely meaningless schedules are still all too evident.

What was the reason for this failure? That's simple. They failed to use the most basic system of all: common sense. They used fancy techniques because they were there, not because they were the correct ones to use to solve their particular problems. They failed to attack their problems logically.

What is so remarkable about material requirements planning (MRP) is that it employs such simple, yet utterly profound, logic. Still, it's no wonder MRP took so long to become widely used: It is not fancy. It does not involve complex algorithms, so it was long considered entirely too pedestrian to merit serious concern by most theoreticians and practitioners. Closer examination, however, reveals its value. For example, what is a master production schedule other than an expedite list—before the fact? And what is a material requirements plan other than a hot list—before the fact? The real difference is how you plan, control, and adjust your priorities, which is why the computer plays such a vital role. Schedule changes can be made again and again with a minimum of difficulty.

Consider the familiar control boards that were standard equipment in production control offices some years back. They were given the generic name of Gantt charts. They fell into disuse because they had to be changed manually and could not be easily kept up. Nevertheless, the logic behind them was valid, for today's production schedule is really nothing more than a computerized Gantt chart.

The point of all this is that modern materials management is not really modern at all, but rather a handling of old concepts the way they should have been handled in the first place. And the key to everything is priority planning.

The failure of the system described above was also caused by what was *not* done. It is evident that capacity projections in conjunction with production planning was not even being considered, with the result that order queues were piling up behind work stations and there was a growing work-in-process in downstream operations. It was precisely this type of situation that gave rise to the need for an expediting function: someone responsible for working through the maze and pulling parts through that were really needed.

## MRP AND STATISTICAL CONTROLS

For too many individuals, to repeat, old habits are hard to break. This is particularly true of order point and order quantity systems, which were in vogue at one time—because they were the only systems available. Fortunately, that is no longer the case. Although the three most popular statistical controls—order point, order quantity, and exponential smoothing—are treated in detail in the following section, a brief comparison with MRP is in order.

### Order Point

As is shown in practically every text dealing with inventory, the order point equation is stated as:

$$OP = D_{lt} + SS$$

or, the order point is equal to demand during lead time plus safety stock.

Suppose a given part number has a projected average usage of 20 parts per month and the order point is based on that quantity. Historical lead time is two weeks, and the safety stock is calculated to be 15. Therefore,

$$OP = 20 \times \frac{1}{2} \text{ month} + 15 = 25$$

Yesterday, when inventory records were checked, 70 units of stock were on hand. Today a stock requisition is received for 60 parts. This drops the on-hand balance to 10 and trips the order point because it is now below the minimum of 25. So an order is placed to bring the stock on hand up to the maximum (say, 75).

You know from past production runs that this part is run quarterly because three months' requirements are run at one time (i.e., $20 \times 3 = 60$). What is needed instead is to bring these parts in or fabricate them when they are required, not when they have dropped below some arbitrary standard. To bring inventory in at this point not only increases total investment but subjects these parts to shrinkage. The impact of this on total inventory investment becomes substantial when this system is extended to cover many of the stocked parts. Under MRP the display looks like Figure 2–2.

## Order Quantity

For this example, assume the part to be a subassembly component of an end item, with the part itself made up of five lower level components. The partial bill of material would look something like Figure 2–3. The column on the left refers to inventory balances as they would likely appear under an EOQ system, while the right column refers to an MRP system.

As is evident from this comparison, EOQ is a perpetual system, whereby replenishments are triggered by order points, while MRP merely reflects residual balances from the last production run. Of more importance, however, is the inventory investment involved. With MRP it is almost negligible; with EOQ, substantial. Moreover, this condition is not likely to change. Balances may go up and down, but total investment is not likely to vary to any great extent. And finally, it is worth noting the patchwork balances that will tend to occur under an order-point, order-quantity system. They happen randomly and bear a relationship to absolutely nothing.

**FIGURE 2–2**
**MRP/Order Point Comparison**

| | | 14 | 15 | 16 | 17 | 18 |
|---|---|---|---|---|---|---|
| Gross Requirements | | | 60 | | | |
| Scheduled Receipts | | | | | | |
| Projected on Hand | 70 | 70 | 10 | 10 | 10 | 10 |
| Planned Order Release | | | | | | |

**FIGURE 2–3**
**Stock Balances: EOQ versus MRP**

| *EOQ Subassembly X* | | *Quantity per Assembly* | *MRP Subassembly X* | |
|---|---|---|---|---|
| *Component* | *No.* | | *Component* | *No.* |
| 1 | 286 | 2 | 1 | 22 |
| 2 | 627* | 4 | 2 | 25 |
| 3 | 34 | 1 | 3 | 7 |
| 4 | 330 | 1 | 4 | 31 |
| 5 | 1497† | 4 | 5 | 1497† |

*order point 150; order quantity of 500 just received
†common-usage part

## Exponential Smoothing

This term simply means *weighted moving average*. Forecasting by this method, and for any other method for that matter, can produce, at best, only satisfactory results, but they will suit your purposes so long as their shortcomings are realized and they are used in the way they were intended. Here, forecasting provides the basis for the time-phased order point system for end item projections (e.g., finished goods).

As a matter of fact, exponential smoothing forecasts of end items are frequently fitted into an MRP format and termed *Distribution Requirement Planning* (DRP) (see Figure 2–4). In this case gross requirements are the forecast amounts that are sat-

**FIGURE 2–4**
**MRP Format: DRP**

| | | 14 | 15 | 16 | 17 | 18 |
|---|---|---|---|---|---|---|
| Gross Requirements | | 10 | 10 | 10 | 10 | 10 |
| Scheduled Receipts | | | 50 | | | |
| Projected on Hand | 20 | 10 | 50 | 40 | 30 | 20 |
| Planned Order Release | | | | | 50 | |

isfied through a normal planning sequence. This plans the arrival of finished goods at a destination or distribution point, like a warehouse.

Below the end item level, however, forecasting should never be used to plan requirements, except in those rare cases where special considerations make it a useful tool—and these are narrowly defined situations, usually beyond your control. But in a one-on-one comparison with MRP, it comes out a very poor second.

## SUMMARY

It is evident that statistical ordering techniques do not belong in a time-phased MRP system because they do not add anything positive. Many practitioners grew up with or inherited such systems and were reluctant to institute any drastic changes. Old habits do die hard!

# CHAPTER 3

# SCIENTIFIC INVENTORY MANAGEMENT

Statistical inventory control techniques are dead and should be buried. They should be expunged from current literature with only a footnote to remember them by. This includes order point[1] and the economic order quantity (EOQ) and all their variations, permutations, and combinations. They are not really *outmoded;* they never were "moded" in the first place. These techniques are based entirely on the following assumptions:

1. They can only be used for independent-demand items. No one, to the author's knowledge, has developed a workable equation for handling dependent-demand items (although I'll bet there's one around somewhere).
2. There is constant usage; they cannot handle "lumpy" demand.
3. There is a steady level of use.
4. The future will resemble the past.
5. There is a certainty of demand; that is, the steady use will continue for the foreseeable future.
6. There is a continuity of demand—it will continue from period to period with no breaks in between.
7. You always adjust formula values for changing conditions—once calculated, the values must be continually updated to reflect current conditions.

---

[1]This does not refer to the the time-phased order point, which performs a very important function.

8. The values assigned to the equation symbols are known—as will be shown, they are not, never have been, never will be.

My challenge to you is this: Go through your inventory records and find those that satisfy these eight assumptions and for these few, if any, you can readily apply the formula.

## INVENTORY REDUCTION

More attention has been paid to the subject of inventory than to any other area of manufacturing. Of course it is crucial to the production process and merits this concern. An old boss of mine used to say that the control of inventory can be reduced to the answers to two simple two-part questions: "What have we got, and where is it?" and, "What do we need, and when are we going to get it?" It's just as hard to argue with his approach today as it was then because the same problems still exist—only the tools to handle them are different.

While my boss's two questions are logical and accurate, they are at the same time a little simplistic; moreover, they make no provision for control. The real problems of inventory and its control are embodied in a different set of questions, which seem to be contradictory: How much inventory must we carry in order to adequately service our customers? How little inventory can we get by with? Over the years, countless techniques have been devised in an attempt to strike a balance between these two contrasting needs, and most of them have failed. The most common ones will now be examined to see why this is so.

## ECONOMIC ORDER QUANTITY

The *theory* behind the economic order quantity (EOQ) is a good one. Too bad it doesn't work in the real world of manufacturing. The EOQ is a square-root formula that balances the cost of ordering, or setup cost, with the cost of carrying inventory. Basically, it tells when it is more economical to incur the cost of a new order, or new setup, compared to carrying a larger lot size over a longer period.

Figure 3–1 depicts carrying costs increasing as inventory accumulates (actually a stepped progression, not a straight line), while unit costs decrease as more parts are run over a common setup. At the point where these lines intersect, you achieve the least total cost. The cost curve is relatively flat over a given range and allows quantity adjustments based on judgment calls. The formula illustrated in Figure 3–1 is

$$a \text{ (carrying costs)} + b \text{ (setup costs)} = c \text{ (total cost)}$$

Eons ago, in a stroke of genius, someone took all these factors and kneaded them into a formula to identify the least total cost, or most economic order, and called it the economic order quantity formula:

**FIGURE 3–1**
**Total Cost Curve**

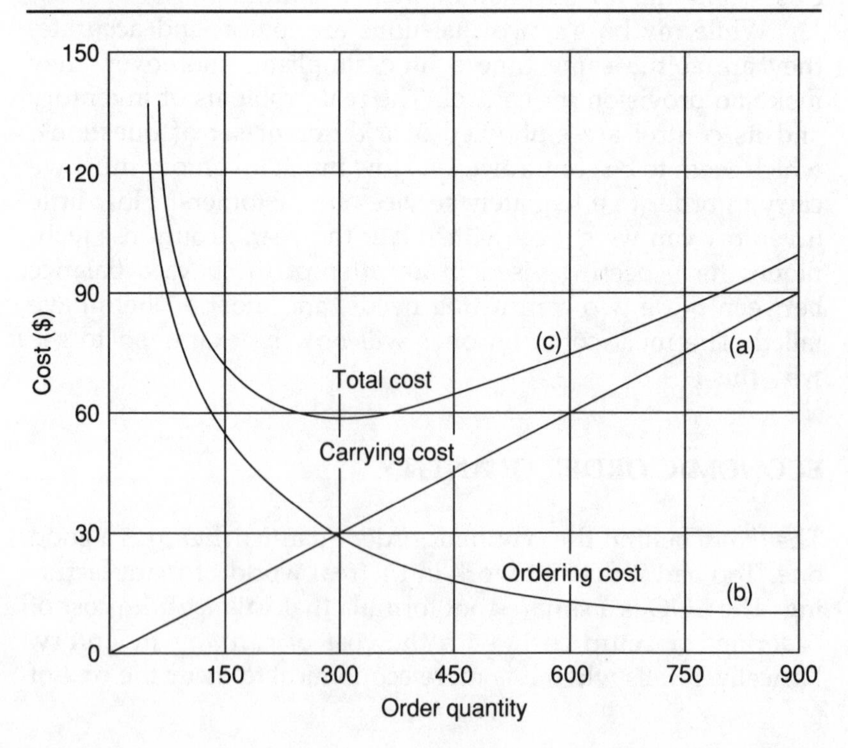

$$EOQ = \sqrt{\frac{2US}{IC}}$$

where

$U$ = The yearly usage of a given part expressed in units

$S$ = The setup, or ordering, cost (cost of acquisition) for this part in dollars

$C$ = The unit cost for the part

$I$ = The inventory carrying charge assigned to the part, expressed as a percentage

An example will show how this works. Assume the usage ($U$) to be 2,000 parts per year at $2 each ($C$). The setup cost ($S$) is $25, and the inventory carrying cost ($I$) is 20 percent.

$$EOQ = \sqrt{\frac{2US}{IC}} = \sqrt{\frac{2 \times 2000 \times 25}{2 \times .20}} = \sqrt{\frac{100{,}000}{.4}}$$
$$= \sqrt{250{,}000} = 500$$

Several things are evident in this rather elementary example. First, in real-life situations inputs and outputs seldom yield such a precise EOQ. This is of no consequence, however, because a computer can crank out an answer regardless of the equation's complexity. Second, the guts of the equation, $S$ and $I$, assume that hard-and-fast numbers can be assigned to them—but they cannot. (Why this is so will be taken up shortly.) Third, calculating the EOQ is simply a mathematical function, and as such is unrelated to the reasonableness of the inputs to the equation.

What may happen can be shown in several examples. Assume that the quarterly forecast, or backlog, is 3,250 units, which equals an annual usage ($U$) of 13,000 units. This is an expensive part (e.g., a motor) with a unit cost ($C$) of $250. It is a purchased part with an $S$ value of $30 and a ($I$) value of 24 percent. Therefore,

$$EOQ = \sqrt{\frac{2US}{IC}} = \sqrt{\frac{2 \times 13{,}000 \times 30}{250 \times .24}} = \sqrt{\frac{780{,}000}{60}}$$
$$= \sqrt{13{,}000} = 114$$

The square root—and therefore the order quantity—is 114 (rounded downward).

The next example assumes a manufactured part for the same end item with the $S$ and the $C$ transposed. This will result in a rather expensive setup cost of $250, a unit cost of $30, and an EOQ of 950:

$$EOQ = \sqrt{\frac{2US}{IC}} = \sqrt{\frac{2 \times 13{,}000 \times 250}{.24 \times 30}} = \sqrt{\frac{6{,}500{,}000}{7.2}}$$

$$= \sqrt{902{,}778} = 950$$

How are quantities 114 and 950 related? Obviously not at all, except that they are common to the same end item, which must be available in a sufficient quantity to meet the final stage of the build schedule. However, this example does point out graphically what can happen when parts are planned from a bill of material independently of each other: Problems result!

Furthermore, consider your planned periodic requirements. How do these EOQs fit in with any logical time frame? With a monthly requirement of 1,000–1,100 (assuming stable, continuous demand), 114 motors isn't even half a week, while the quantity 950 is the better part of a month's run. In cases like this you need to make a judgment call and change the quantity to one you feel more comfortable with. *This is exactly what the EOQ was intended to preclude!*

Finally, what shift supervisor or plant foreman, when considering the high cost of setup and experiencing a smooth production run, would not just use the scheduled EOQ arbitrarily and run up the entire quarterly requirement. Just think of the efficiency!

## UNKNOWNS

### Setup Costs

$S$ in the equation stands for *setup cost* if the part is to be manufactured in-house, *ordering cost* if purchased from an outside vendor. The two obviously present entirely different pictures from a cost standpoint. The first setup cost to consider is the cost of wages: the hourly wage of the setup person times the number of hours allowed for setup. Experience will show that this is likely to be different each time the machine is changed to run this part. Are these times to be averaged out? And does this do justice to a minor setup like a simple die-switching, as compared to a complete changeover? Due to the independent status of the part being manufactured, you cannot know which group of parts it is likely to run with, unless you plan it that way. However, grouping families of parts will rarely be the same number as the EOQ. What is usually done in practice is to take the easy way out and use the standard hours you get from Engineering or Accounting. After all, these are always accurate!

Other labor costs present problems too. Whereas wage rates are a fairly straightforward number, accounting departments do not settle for these alone in their inclusion of standard costs. They want to cover the entire waterfront of costs—known and unknown, real and imaginary costs lumped together into a package labeled *burden,* or some equally official-sounding term whose relationship to actual costs is merely assumed. For example, a labor rate of $5 per hour carries a burden rate of $5, and a $10 rate carries a $10 burden, thereby projecting a linear relationship between the two with little thought about its validity. It is a convenience.

The two main components of the setup cost, the labor rate and the burden rate, are now in place. Adding the two will give you the per hour cost of making most setups. Engineering has calculated how many hours it takes for any particular setup; so the cost is calculated by multiplying the labor rate by the number of hours allowed. It should be noted, however, that the hours allowed and the hours actually used to complete the setup are the same only by coincidence. It is entirely possible that a 4-hour setup could just as easily take 3 hours or 6 hours, or some other number entirely different. The point is, just as it was with the burden rate, the setup cost is just a number whose validity is merely assumed.

This is not all. One must also consider what some refer to as the *theory of alternate use* (or *investment*) *or opportunity costs.* In other words, while a machine is being set up, it cannot run production, and vice versa. This suggests that included in the setup is the value of the components that could have been run. If so, what costs are to be picked? Whatever costs you choose are artificial. In the final analysis, the machine does have to be changed over, but you should bear in mind that employing the alternative-use technique will only serve to greatly inflate the calculated quantity.

Similar arguments can be made regarding ordering costs. Suffice it to say that identifying and summing up the bits and pieces you feel should be included in the cost of ordering can be carried to ludicrous extremes—all precise to three decimal places. Here again, any relationship between these estimates and what it actually costs to place an order can only be guessed at. They cannot be sufficiently quantified.

**Inventory Carrying Charges**

The *I* in the EOQ formula is the truly blue-sky part, because some users believe the figure should be 10 percent, while others say 40 percent is more accurate. Understandably, this can cause serious problems, because calculating the formula using such widely divergent values will produce widely divergent EOQs. This is also a tipoff that something isn't quite right with the formula itself.

Returning to the first example, annual use is still 2,000, costing $2 each; and the setup cost remains at $25, but now *I* is set at 10 percent.

$$EOQ = \sqrt{\frac{2US}{IC}} = \sqrt{\frac{2 \times 2{,}000 \times 25}{.10 \times 2}} = \sqrt{\frac{100{,}000}{.2}} = 707$$

Compare this result with what it would be if *I* were raised to 40 percent:

$$EOQ = \sqrt{\frac{2US}{IC}} = \sqrt{\frac{2 \times 2{,}000 \times 25}{.40 \times 2}} = \sqrt{\frac{100{,}000}{.8}} = .354$$

This clearly illustrates the problem. Which quantity is to become the lot size: 707? 354? somewhere in between—like the original 500? or none of the above? Exactly which point best reflects the conditions existing in your company? The answer is, nobody really knows—they guess. Furthermore, the climate in which the business operates is constantly changing. Is the inventory carrying cost ($I$) periodically revised to take into account those changes? Here again, the answer is probably no. If the carrying-charge percentage was originally suspect, at best, it certainly is not likely to improve with time.

Here are some of the common components of inventory carrying costs (ICC):

Obsolescence.
Deterioration.
Inventory taxes.
Inventory insurance.
Storage costs.
Cost of capital.

The troubling thing about these six categories (eight, if you include out-of-stock and capacity-related costs) is that they each have a blank that must be filled in. The question then becomes how much to allocate to each factor. Then you throw in enough to make sure it's covered, because that is the scientific way.

The question also arises whether they are legitimate. Because it would be too time-consuming to list the good and bad points of each, just one component will be examined: storage costs.

Calculating storage costs would seem to be pretty straightforward: Multiply the square footage of the space designated for storage by a per-square-foot cost and you have the total storage cost. Unfortunately this is too simplistic and begs too many questions.

For example, assume you have a 100,000-square-foot building that costs $10,000 per month (lease or buy—it doesn't matter). Of this space, half, or 50,000 square feet, is designated as warehouse, or storage space, so you will have an annual cost of $60,000 (12 × $5,000). The first problem you must face is this: When you entered into a contract to lease, buy, or build, you legally encumbered yourself to pay X dollars over Y months. It doesn't matter whether you use 1 square foot or the entire 100,000, you will have to pay $10,000 each month just the same.

Another question you must consider is this: Are you going to charge off storage at the same rate regardless of the value of the items stored? Certainly finished goods have a higher value and take up less space than the components they are made from. And certain raw materials, like electronic components and stainless steel, obviously have more value per square foot than packaging supplies or other bulky items. Furthermore, is the storage rate to be continually adjusted as the amount of space used over a whole range of components ebbs and flows? Hardly.

In the same vein, suppose there is excess storage capacity from time to time. The Accounting Department will tell you to include all available space into carrying costs because if you are not using it, it could be rented out. They are dealing in pure fantasy. Whom are you going to rent it to? How would you go about finding some idiot who would inconvenience himself by renting out a few feet of your floor space on a temporary basis at your stated rate when he could get all the space he needs at a public warehouse and probably save money in the bargain? The rationale may seem to be logical, but it certainly doesn't apply in the real world.

This could go on and on, but the point has been made. The crux of the matter is this: The cost of maintaining a storage facility is tied more to the cost of doing business than it is to any carrying charges. You are paying for the facility each month regardless of how you massage the numbers. It is no less a cost of being in business than is the inventory itself. The same thing is true of taxes, interest, etc. These things are inherent in operating a business. If they can be quantified, they belong in the "cost of goods sold" section of the income statement. Period. To pile a lot of vague nothingness into inventory carrying cost in the EOQ formula because there is a blank to be filled doesn't make a lot of sense. A complete list of the unknowns in the EOQ is shown in Figure 3–2.

This is not to suggest that there are no costs associated with carrying inventory. There are—and probably in real dollars. But because there is no way of knowing precisely what the

**FIGURE 3–2**
**Unknowns in the EOQ**

Setup costs
- Burden
- Overhead
- Opportunity costs

Ordering costs
- Clerical and paper-work-associated costs

Unit cost

Inventory carrying costs
- Obsolescence
- Deterioration
- Insurance
- Storage

Out-of-stock costs

Capacity-associated costs
- Hiring
- Firing
- Training
- Overtime
- Idle time

costs are, you have no choice but to ignore them. Even if this were not true and these costs could somehow be quantified, what would be accomplished? The EOQ is intended to be an instrument for the control of inventory. Not only does it not do this, but its effects are exactly the opposite. However, the cause of this failing is only to a small degree due to the fact that equation symbols are impossible to quantify. No, the problem goes a lot deeper than that.

A survey concerning the value of EOQ was conducted some years ago, and it revealed that fully 80 percent of the respondents found no discernible difference in the level of their inventory employing the EOQ technique as compared to the system they had previously used. Of those who did report a change, more than half indicated that their inventory actually rose. For the remaining few who said their inventory decreased, the reduction probably had more to do with management's mandates than it did through order quantity manipulations. In any case, none of this is surprising, for the order quantity formula says nothing about lowering the level of inventory, only that the order quantity be "economic."

### Unit Costs

Although the setup costs ($S$) and the inventory carrying costs ($I$) have been singled out for scrutiny, it should be pointed out that the charges of vagueness and obscurity apply equally well to the unit cost ($C$). The standard cost supplied by Accounting is no more accurate than were those associated with setup costs. It is merely a number, and it is generally recognized to be fictitious.

Finally, let's devise a new version of the formula that more accurately describes EOQ:

$$EOQ = \sqrt{\frac{2UE}{TG}}$$

in which $U$ is the same as in the original equation, $E$ is an *estimate* of what you think setup costs are, $T$ is a *toss-up* as far as unit costs are concerned, and $G$, for carrying charges, is a pure *guess*.

Now you can insert numbers that you are convinced exactly represent the symbols and compute an EOQ precise to three decimal places. What is puzzling is how otherwise intelligent and successful business people can hold with such nonsense.

## MANAGING INVENTORY LEVELS THROUGH LOT-SIZING RULES

Since MRP is basically an inventory management system, a great deal of attention has been paid to lot-sizing techniques in an effort to get a handle on schedule requirements. Lot-sizing rules are discussed here rather than in the section on MRP because many of the techniques employ some sort of least-cost principle (using the $E$ and $G$ factors from the formula above), and knowing which are viable will facilitate the discussion later on. Some common lot-sizing rules are:

ABC analysis.
Economic order quantity.
Fixed-order quantity.
Fixed-period quantity.
Least total cost.
Least unit cost.
Period order quantity.
Part-period balancing.
Lot-for-lot.

Two other lot-sizing rules, McLaren's Order Moment and the Wagner-Whitin algorithm are too off the wall to merit inclusion here. Besides, these are both variations of part-period balancing and, as such, do not make use of common sense. Each of the concepts is briefly discussed along with the relative merits and demerits of that particular technique.

### *ABC Analysis*

Also commonly termed *Pareto's law* or the *80/20 rule* and more formally known as *distribution by value.* Basically, it states that 80 percent of any particular set of occurrences result from 20 percent of their causes. In Pareto's law 80 percent of Italian wealth was held by 20 percent of the families. In this case 80 percent of inventory dollars result from 20 percent of the parts. This is the most rudimentary of all inventory management concepts. It really doesn't "do" anything except break down a mass of data into logical categories. Due to its simplicity, however, it provides a useful concept in a great many instances.

### *Economic Order Quantity*

The less said of this, the better.

### *Fixed-Order Quantity*

Whenever a quantity is ordered or run, it is always a fixed amount. Whenever there is a constraint (heat treat oven size) or a benefit to be gained (cash discount, truck load quantities), this approach may make good sense. Figure 3–3 shows an example.

**FIGURE 3–3**
**Fixed-Order Quantity**

| | 14 | 15 | 16 | 17 | 18 |
|---|---|---|---|---|---|
| Planned Coverage | 100 | 100 | | | 100 |

**FIGURE 3–4**
**Fixed-Period Quantity**

| | 14 | 15 | 16 | 17 | 18 |
|---|---|---|---|---|---|
| Planned Coverage | 75 | | 110 | | 95 |

### *Fixed-Period Quantity*

Similar to above, but the time sequence for ordering is specified and the quantity may vary from order to order, or whenever ordered, the quantity will cover a fixed number of periods. Again, if it fits your needs, use it. This technique is shown in Figure 3–4.

### *Least Total Cost and Least Unit Cost*

These techniques are similar in that they are based on the least-cost principle embodied in the graph in Figure 3–1. As carrying costs rise in some relationship to increases in quantity, setup costs go down because they are spread over increasingly more units (called *economy of scale*). In theory, where the two graph lines intersect is the point of optimum value because costs are in balance. But because the constraints are the "estimate" and "guess" factors (i.e., setup costs and ICC), the resulting quantity calculated is merely an assumption. This presents a good academic exercise and can only be used for that purpose.

### *Period Order Quantity*

Many texts present a formula to calculate this quantity. Forget it. Use common sense. All too many of the formulas that have sprung forth relative to lot size computation are based on assumptions that may or may not be valid. This one is no different. The only rational way to apply the period order quantity (POQ) is to scan your planning horizon and mentally group two or more upcoming requirements. See if it looks sensible to include them into one or several orders, and follow your intuition. Completely dismiss the fact that ICC charges may be incurred for holding this inventory an extra period or so. These could even be regarded as safety stock. Contrary to some beliefs, there is yet to be developed an equation that transcends common sense. The POQ is depicted in Figure 3–5.

### *Part-Period Balancing*

One of the weaknesses of POQ as calculated by its equation made it necessary to carry inventory through periods of low, discontinuous (nonuniform) demand. To get around this problem, part-period balancing was devised.

First, one calculates the economic part-period factor (EPPF). Using the EPPF as the base number, successive periods of demand are plotted against the EPPF until it fails (i.e., will cover no more periods). This feature is called the *look-ahead.* If these future periods over- or under-utilize the quantity of the EPPF, periods of planned coverage are reviewed to see if reducing previous orders would be more effective. This is called *lookback.* Obviously, reducing previous orders entails additional setups. So part-period balancing tries to balance the cost of carrying inventory through successive periods with the cost of making another setup. This is nearly identical to the least-total-cost approach, to which it is directly related. Part-period balancing fails on three accounts:

1. A closer look at the EPPF shows it to be nothing more than the traditional EOQ treated somewhat differently.
2. EPPF entails great computational simulation, and the results are suspect at best.
3. Schedule changes render all this effort useless.

**FIGURE 3–5**
**Period Order Quantity**

| | 14 | 15 | 16 | 17 | 18 |
|---|---|---|---|---|---|
| Net Requirements | 35 | 10 | | 40 | |
| Planned Coverage | 85 | | | | |

### *Lot-for-Lot*

Simple, straightforward, and easy to use, this technique does not require computing useless formulas. Planned order quantities are generated exactly equal to the net requirements for the period in which they are scheduled. This is a prime example of what is meant by common sense manufacturing. Critics say it is a main cause of system "nervousness," especially when brought on by schedule changes necessitating order replanning. But those who think contrived order quantities, particularly those that employ the *E* and *G* factors, provide more security are only kidding themselves. It is not so much the formula that drives a given lot-sizing rule that reduces this nervousness as the higher level of inventory that it promotes. Application of lot-for-lot is shown in Figure 3–6.

Some authors describe lot-sizing categories as being either demand-rate-oriented or discrete—the latter meaning ordering to the extent of net requirements with no remnants to carry forward for periods in which there are, as yet, no current requirements.

What lies at the heart of the matter, however, and more important than mere classification is the primary objective: the level of customer service. The choice of a lot-sizing rule must provide a reasonable solution to the two contradictory questions posed at the beginning: How much inventory must be carried to ensure meeting customer requirements? And how little can that be? One's selection of a lot-sizing rule will directly impact the level of inventory investment. Equation-driven techniques that are demand-rate-oriented pay little heed to the neutralizing of these opposing questions. The answer seems to favor the selection of lot-for-lot, for that satisfies the requirement need while keeping inventories at respectable levels.

**FIGURE 3–6**
**Lot-for-Lot**

| | 14 | 15 | 16 | 17 | 18 |
|---|---|---|---|---|---|
| Net Requirements | 35 | 10 | | 40 | |
| Planned Coverage | 35 | 10 | | 40 | |

Of course lot-for-lot should be tempered with period order quantity in order to reap any benefits that might exist. Other lot-sizing rules (e.g., fixed-order quantity) are merely a reflection of the specific environment in which your company operates—assuming, as always, no negative impact on customer service.

## ORDER POINT

The order point system, which has served so unfaithfully in the past, has given way over recent years (albeit sometimes reluctantly) to material requirements planning (MRP). Despite some claimed advantages, it has enough glaring weaknesses to render it useless for any practical consideration.[2] MRP is just the opposite. The reason MRP sometimes fails to deliver the "advertised" advantages will be taken up in a later section.

The logic behind the order point is easy to understand. Its purpose is to ensure that part inventories will be available when called upon. Traditionally this was done when the quantity reached its reorder point. Using statistical techniques to forecast the rate of demand, the quantity that would be consumed during the planned lead time necessary to replenish stock was set as the reorder point. As protection against forecast errors, an amount of safety stock was added to this reorder point. Then, as stock was consumed and a new balance calculated, it was checked against the reorder point. When this point was reached, it triggered a replenishment order to come into stock, depending upon the length of lead time, just as the remaining stock on hand was used up. The formula is written

$$OP = D_{lt} + SS$$

as was mentioned in the preceding chapter. The order point is equal to demand during lead time plus safety stock. To put it another way: If you know how long it takes for an order to be delivered to stock and you can accurately estimate how much

[2] Again, this is not to be confused with the time-phased order point.

you will consume during this intervening period, you will know the demand during lead time. Safety stock covers those occasions when demand or usage exceeds the estimated demand.

When the stock on hand drops below this calculated order point, it is a signal to reorder. The quantity of the order is determined by the EOQ. For example, a user wishes to calculate the order (or reorder) point for a given part. Stock history shows that it takes four weeks to get delivery—this is lead time. It's estimated that during the four-week period 200 units of this part (50 per week) will be used—this is demand. Safety stock is some quantity over and above this, usually an arbitrary percentage of the order point—for "just in case." In this example

$$\begin{aligned} OP &= D_{lt} + SS \\ &= 200 + 25\%\ (200) \\ &= 200 + 50 \\ &= 250 \end{aligned}$$

When the balance of the stock on hand drops to this point, the planner will reorder up to a maximum, which number is the EOQ. If this number is 500 and the order point is 250, the maximum total inventory is 750. This type of inventory control is known as the *min/max technique*. It failed for one very practical reason. While the order point system answered two of the three questions needed for inventory control telling how much and when to order—the third question (when it was needed) wasn't resolved until material requirements planning arrived on the scene. The functioning of the order point is generally visualized by the sawtooth diagram in Figure 3–7.

The final indictment of scientific inventory management, in relation to the order point system and EOQ, can be seen by examining data from actual field conditions. Suppose a user wishes to build a component called Assembly C, which is made from five different parts. The order calls for 100 assemblies, and the inventory computer printout shows the following stock on hand.

**FIGURE 3–7**
**Sawtooth Diagram**

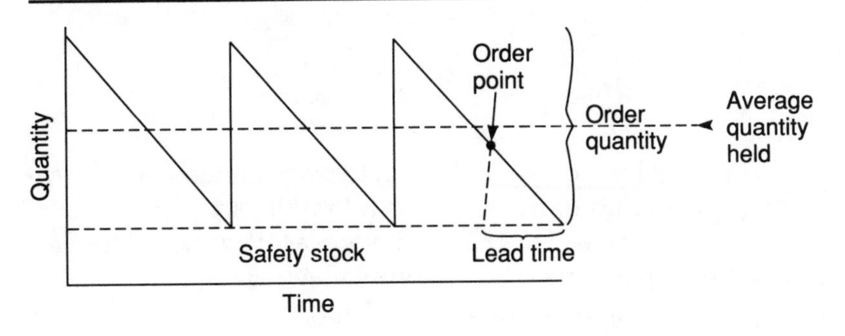


*Assembly C (Gross requirement 100)*

| | |
|---|---|
| Part 1 | 880 |
| Part 2 | 350 |
| Part 3 | 400 |
| Part 4 | 50 |
| Part 5 | 620 |

The problem facing the user here is to ascertain how many completed assemblies can be built with this component inventory? The fallacy of the statistical inventory techniques is obvious immediately. With it one orders each part independently, without regard to its relationship to the other parts that must also be available to complete the order. Thus, even with a total inventory at an unusually high level, one can only build 50 units of Assembly C.

Look at the same problem from another perspective. Suppose your management mandates that you keep a minimum inventory 90 percent of the time. In other words, if you go to pull a part 10 times, it will be available no less than 9 of those times. With this system, will the necessary materials be available when required most of the time? The answer is: not necessarily. In fact, statistical evidence shows this to be highly improbable. Consider the following example:

*Assembly C (Gross requirement 100)*

| | |
|---|---|
| Part 1 | 90% availability |
| Part 2 | 90% availability |
| Part 3 | 90% availability |
| Part 4 | 90% availability |
| Part 5 | 90% availability |

When stocking personnel begin to stage material for an upcoming run and go to the stockroom to pull Part 1, they are 90 percent sure that it will be there. The process is repeated for Part 2, which is 90 percent certain to be available. However, what is most important, in order to satisfy a build schedule, is that Parts 1 and 2 must *both* be available at the same time. The law of statistical probability says that because each part has its own 90 percent availability, the probability of both being there at the same time is 90 percent times 90 percent, or only 81 percent.

Likewise, Part 3 is available 90 percent of the time, so the law of probability holds that the chances of all three parts being on hand simultaneously is .9 (90%) × .9 × .9, or 72.9 percent. When this process is repeated for all five parts (see Figure 3–8), one finds the likelihood is that Assembly C can be built only a little more than half the time (59%). In fact, if the assembly has 10 parts, each with 90 percent availability, the chances of having all 10 parts in stock is barely one in three (35%). Happily, all is not lost. There are still expeditors, hot lists, and safety stocks.

**FIGURE 3–8**
**Statistical Part Probability**

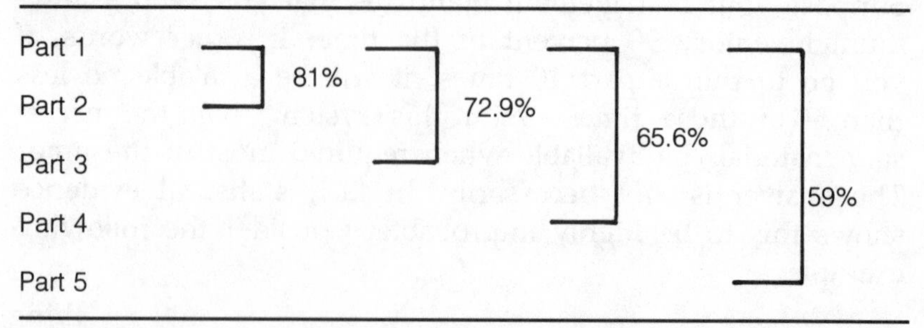

In any case, this is not what is meant by inventory control. The net result of all this activity is inventory accumulation. And because of the assumptions that are part of all equations, the need for safety stock is an integral part. Take another look at the formula. $D_{lt}$ in the formula merely asks, "How much are you going to use (or sell) before you get a replenishment?" Then the order point is calculated and a replenishment is ordered with the purpose of bringing the inventory level *up to* a given point—irrespective of whether there are any current requirements.

## FORECASTING

Although the next chapter is devoted to forecasting, it should be mentioned here as it relates to statistical inventory management. Using forecasting for determining component part inventories automatically makes them subject to many of the assumptions that make the EOQ and the order point invalid. Forecasting, while extremely important, is only applicable in situations like projecting the level of end items (e.g., finished goods). Below the end item level, however, forecasting should never be used to plan requirements, except where special considerations make it a useful tool. These narrowly defined situations, generally beyond one's control, will be described in detail later.

One bit of wisdom concerning forecasting should always be kept in mind. Forecasting is an imperfect science and its use should be restricted to those conditions and circumstances where better controls are not available. With reference to planning lower level requirements using forecasting techniques, however, better controls do exist.

## SUMMARY

The dismal record of the three systems of inventory control becomes apparent when they are examined closely. Forecasting, through exponential smoothing, does not add anything positive. By their very construction, the statistical inventory management techniques of the EOQ and the Order Point actively promote the accumulation of inventory—not the reverse. These formulas have never saved as much as a single dollar of inventory, and they never will.

# PART 3

# MATERIALS MANAGEMENT

## OVERVIEW

The American Production and Inventory Control Society (APICS) defines *materials management* as "the grouping of management functions supporting the complete cycle of material flow, from the purchase and internal control of production materials to the planning and control of work in process to the warehousing, shipping, and distribution of the finished product."[1]

This definition does not quite fill the bill, but it seems to be more appropriate for the purposes here than some of the more recent developments, chief among these being "closed-loop MRP" and Manufacturing Resource Planning. Because these terms are similar (yet not identical), both are commonly referred to as MRP II, which suggests something more comprehensive than the planning of requirements.

Closing the loop, as the former term states, means taking those functions involved in the making of a product and fitting them together to form an integrated planning and control

[1]American Production and Inventory Control Society, *APICS Dictionary*, 6th ed. (Falls Church, Va.: APICS, 1987), p. 8.

system, with the completion of one process leading directly into the next and so on. Manufacturing Resource Planning means to treat the manufacturing organization as a unified whole focusing more on goals and objectives than on the steps necessary to achieve them.

Both of these systems (actually one is just an extension of the other) are too company focused. Materials Management is material oriented; closed-loop MRP is more process oriented. Materials Management implies service to the customer while closed-loop MRP, or Manufacturing Resource Planning, refers to it only vaguely.

The purpose remains, as it has been throughout, to improve the level of customer service. The way this is accomplished is secondary. Treating the subject from the standpoint of managing the materials seems to be a more effective way of achieving this purpose.

I would like to make the definition of Materials Management more complete by adding at the end "finished product to fill a customer need."

A similar definition of materials management is offered by the National Association of Purchasing Management: "A confereracy of traditional material activities bound by a common idea—the idea of an integrated management approach to planning, acquisition, conversion, flow, and distribution of production materials from the raw-material state to the finished-product state." A similar statement regarding the integration of customer needs with this definition is likewise suggested here.

# CHAPTER 4

# FORECASTING

Forecasting is an important tool in the materials manager's arsenal, and it is difficult to see the department functioning efficiently without it being actively pursued. Although forecasting is used in other functions (primarily marketing and sales), the concern here is to explain its uses and detail the various techniques that are of most help in production and inventory planning and control.

Statistical forecasting through the analysis of masses of data, as done by various trade associations and government agencies, involves very complex methods. Among the best known are the techniques of least-squares, leading indicators, regression analysis, and correlation. These are not of concern here, although if you have a background in statistics and you find these methods useful in solving thorny problems, by all means use them. But by and large, we will employ only the most elementary arithmetical statistics—and a large dose of common sense—to solve most planning and control problems.

## LAWS OF FORECASTING

Before examining forecasting in detail, it is important to understand the three basic laws of forecasting:

1. All forecasts, by definition, are bound to be wrong. Forecasts involve predicting the future based on an analysis of historical data. The assumption is that the

future will probably be like the past, and it is through the use of historical data that one attempts to predict the degree of forecast error.

2. Forecasts of product groups are far more reliable than forecasts of its individual members. As the group increases in size, forecast error decreases. Forecasts of individual members may vary widely from actual fact, but their total should approximate what actually happens.
3. The shorter the forecast period, the more accurate it will be. Understandably, when you try to predict future events from circumstances as they stand, the farther you project into the future, the less able you are to ascertain the impact of external forces.

So while one can acknowledge that a forecast is going to be wrong, one wants to know the degree of forecast error. The ability to make sound management decisions is based on the ability to control this factor. In many cases, the forecast number will impact on the rate of production. Inaccurate forecasting can have a negative effect on either side of the inventory coin: poor customer service, from out-of-stock situations, or excessive inventory. In many instances the forecast is the main input to production scheduling.

## TYPES OF FORECASTS

There are, in effect, only two types of forecasts: judgment forecasts and statistical forecasts. They are sometimes referred to as extrinsic and intrinsic forecasts, respectively. Because of the types and availability of data used by the two types of forecasts, they are also called qualitative and quantitative. A list of the various categories of forecasting, as well as the situations under which they might be applied, is presented in Figures 4–1 and 4–2. For the purposes of this discussion, however, only the more popular and simpler types have been singled out for treatment.

**FIGURE 4–1**
**Qualitative Forecast Types**

*Market Research: Indirect Forecasting More Appropriate to Marketing*

Questionnaires
Consumer intention surveys
Focus groups

*Panel Consensus*

Sales force estimates
Jury of executive opinion (Delphi approach)
Survey of customer intentions

*Historical Analogy*

Comparative analysis of past events

*Management Estimation*

Executive opinion applied to detailed forecasts

*Visionary*

Long-range in nature

## Judgment Forecasts

These qualitative forecasts are based on the subjective judgment of individuals. There are three main categories:

1. Jury of executive opinion. This approach combines the judgment of a group of managers—usually the head of each operating department—about the forecast. These judgments are supported by historical data, economic and industry conditions, and competitive actions. Although this approach is the most widely used of the forecasting techniques, its products categories are more general and it covers a longer term, usually a year or more. It can be refined by the Delphi approach, according to which the forecasts are submitted anonymously by each member, discussed by the group, and then the action

**FIGURE 4–2**
**Quantitative Forecasts Types**

*Extrinsic (external data sources)*

Published
- Leading indicators
- Gross national product (GNP)
- Census
- Demographic compilations and breakdowns

Original
- Sales force customer poll

*Intrinsic (internal data sources)*

Time series (from demand history)
- Naive method
- Moving-average methods
  - Single
  - Double
  - Weighted
- Exponential smoothing (form of weighted moving average)
  - First-order (simple)
  - Second-order (trend)
  - Third-order (extra trend, fads, fashions)

Other forecast models
- Base models
  - Seasonal pattern smoothed exponentially
- Trend (seasonal)
  - Seasonality and trend analysis
  - Combined
    - Winter's method
- Adaptive smoothing: adjusting the forecast automatically using a sophisticated form of exponential smoothing
- Box–Jenkins model: complex mathematical correlation model examining demand over certain given time periods
- Trend projections: similar to Trend, above, but looks long-range

*Causal Models*

- Simple regression: trend analysis utilizing least-squares
- Multiple regression: predicting the combined effects of two or more independent variables on a dependent one
- Econometrics*: series of multiple regression equations premised on socioeconomic conditions
- Leading indicators: factors whose movement presages certain events (not actually causal)
- Lagging Indicators*: same as above, except indicators confirm that the event has taken place
- Input-output models*: summarizes effects of all transactions that take place at the macroeconomic level

* Not applicable except in long-range projecting

repeated until a consensus is reached. The purpose here is to dilute any one individual's input to the group effects.

2. Sales force estimates. Each salesperson submits a forecast based upon each product or product group by customer. It is assumed that he or she is closest to the action and will have a more complete and current knowledge of the customer's behavior and intentions. This is another widely used approach, although it usually performs better in industrial-product firms, where there are relatively few customers and/or products. A word of caution: If compensation is based in part on these forecasts, which then become sales quotas, they may become biased.
3. Surveys of customers' intentions. Customers are asked to make their own forecasts about projected usage, because the customer is supposed to have the best knowledge upon which to base a forecast. Here again, the fewer the customers, the better this approach works, so it is of greater use to firms that make industrial products.

**Statistical Forecasts**

For production and inventory control practitioners, statistical forecasting is forecasting for the near term, generally up to three months. Such forecasts are quantitative (unlike judgment forecasts) and are also referred to as intrinsic forecasts or time-series projections. The approach is quite simple: You simply take historical data and extrapolate forward in time.

**Statistical Forecasting Considerations**

In order for such forecasting to be appropriate, a number of considerations must be taken into account. First, and most obvious, historical-demand data must exist. Second, it assumes the future will resemble the past and is relatively stable. Third, it must be possible to detect demand patterns (trends) in past data. Fourth, demand must be unimodal, that is, you cannot have sales of, say, 500 or 1,500 and seldom anything in between. Finally, demand variations are assumed to be symmetrical, that is, deviations from forecast on the plus side and the minus side should tend to be equal or very nearly so.

One problem you may encounter is with very low forecasts. If the forecast is, say, 20, the variation on the plus side might be as much as 100; but on the low side it cannot be lower than zero. One possible solution would be to disregard those periods where demand is zero (or extremely low, provided there are not too many and they are unique) when calculating the averages.

The last consideration combines both qualitative and quantitative techniques. Termed *causal influence,* it refers to variations in demand that seem to be linked to some external factor, which you may or may not be able to identify by monitoring a set of variables. The most popular of these variables is the leading indicator. However, the uses and treatment of the data thus obtained belong more in the field of regression analysis and other state-of-the-art econometric techniques and are beyond the scope of this discussion.

## Statistical Forecasting Tools and Techniques

### *The Moving Average*

Time-series projections use only historical data on the variable to be forecast. The single most powerful of all short-term forecasting tools is the moving average, and other short-term forecasting systems are merely variations of it. Even though the moving average is termed a statistical concept, calculating it requires only the knowledge of simple arithmetic.

Sales figures for actual demand are listed for each period, typically the 12 months in a year. The total for the year is divided by 12 to get a monthly average. The sales for the current (13th) month are then added to the total and the first month dropped. The new total is again divided by 12 to get a new monthly average. This process is repeated for each succeeding month, adding the new sales figure and dropping the first. Each new monthly average will be a reflection of expected current demand at any particular point in the sales year. It should be noted, however, that these new averages are not the forecast numbers that will be used to plan products. Certain refinements must first be factored in, but the new moving average does form the basis for all future forecasting. Figure 4–3 shows this. The next calculation will begin with month 3 and include month 14.

Obviously a weekly forecast using the moving-average method can also be generated, but a week is too short a turn-around time for practical planning purposes. It is further inadvisable to use periods of less than a year for moving-average calculations because a period of six or nine months could negate whatever seasonality is involved. Six months of lower-than-normal activity would understate the forecast as the organization moves into its heavy season; and the converse could also occur.

The moving average, for all its good points, has drawbacks. First, monumental amounts of data need to be stored, especially where there are many items of inventory that must be forecast. Back in the early days of computers when the amount of storage capacity was limited, this presented quite a problem, but with the advent of the modern computers this problem has largely disappeared. Second, as is true of all statistical techniques that are based on historical data, the moving average lags behind actual changes in demand. Random occurrences falling outside the normal range will eventually be picked up in the process of averaging, but trends in demand do not allow for self-correction. (Techniques to detect these changes will be discussed later.)

Finally, one of the considerations governing the appropriateness of statistical forecasting is the assumption that the future will resemble the past and that demand is relatively stable.

**FIGURE 4–3**
**Moving Average**

1 2 3 4 5 6 7 8 9 10 11 12
120 + 150 + 140 + 140 + 125 + 105 + 118 + 50 + 40 + 55 + 60 + 70 = 1,173
1173 ÷ 12 = 98 (base average)

2 3 4 5 6 7 8 9 10 11 12 13
150 + 140 + 140 + 125 + 105 + 118 + 50 + 40 + 55 + 60 + 70 + 92 = 1,145
1145 ÷ 12 = 95 (first-period moving average)

History shows us, however, that this is not always true. Certain external forces can come to bear that can cause wide fluctuations in demand patterns, even in the short term. Two cases in point are the oil crisis of the 1970s and Iraq's invasion of Kuwait in August 1990. In the case of the latter, the degree of stability does not have a direct effect on the number calculated by the moving average, but the relative stability or instability does impact directly on the amount of safety stock that must be carried to cover demand fluctuations. In either case, the value of the moving average should not be judged totally on how well the future resembles the past nor the stability factor but on the basis of whether there is something around that will do the job better.

Applications of one or more of the variations of the moving average may prove to be beneficial, but in the final analysis the result is likely to be roughly the same. After all, no forecasting system yet devised could have forecasted the effects of such events as the oil crises.

### *Exponential Smoothing*

Probably the best-known and most-popular forecasting technique is exponential smoothing. In spite of its pretentious name, it is simply a weighted moving average. Its purpose is to assign a given weight to the most recent sales activity while still assigning a greater percentage of the weight to the old forecast. This is done by applying an alpha factor to the most recent sales data, expressed as either a decimal or a percentage. Several examples will show how this works.

Assume the sales forecast is 100 and the actual sales for two months are 80 and 110.

| | | |
|---|---|---|
| Forecast | 100 | 100* |
| Sales | 80 | 110 |
| Average | 90 | 105 |

*If this were assumed to be the second month, this number would be something less than 100. Using 100 merely serves as a convenient example.

The above is, of course, a simple average. It assumes that an equal weight of 50 percent (.5) is assigned to both numbers. If 50 percent were converted to its decimal equivalent of .5 and calculated exponentially, the equation would look as follows:

| | | |
|---|---|---|
| Forecast | 100 × .5 = 50 | 100 × .5 = 50 |
| Sales | 80 × .5 = 40 | 110 × .5 = 55 |
| New forecast | 90 | 105 |

You may still be convinced that the forecast is more valid than the sales figures for any given month, that it is too high and thereby does not give sufficient weight to the ongoing forecast. If the assumption is made that the old forecast deserves an 80 percent (.8) weight in the calculation of a new forecast, the alpha factor would be its reciprocal, or 20 percent (.2). (Keep in mind that the two figures must always equal 100 percent, or 1.00.) The two decimals, .8 and .2, are now used to calculate an exponentially smoothed forecast:

| | | |
|---|---|---|
| Forecast | 100 × .8 = 80 | 100 × .8 = 80 |
| Sales | 80 × .2 = 16 | 110 × .2 = 22 |
| New forecast | 96 | 102 |

It now becomes clear that the new forecast approximates the planned forecast much more closely than does the simple average. If an alpha factor of .1 (.9 for the old forecast) were chosen, the new forecast would be 98 and 101, respectively.

This raises the question of which alpha to use, which best reflects recent history and experience. The answer to this can be found by employing either of two equations. The first equation works from the number of moving-average periods you want covered in the forecast to the alpha factor to be used. The second equation works from a given alpha factor felt to be representative to the number of moving-average periods to be covered.

$$A = \frac{2}{N + 1}$$

$$N = \frac{2 - A}{A}$$

where

$A$ is the alpha factor
$N$ is the number of periods

Using a computer, a great many simulations can be done to compare the effects of different variables. In the final analysis the alpha factor is most generally set using common sense. The following list shows the most often used values:

**FIGURE 4–4**
**Alpha Factors and Moving-Average Periods**

| *Alpha Factors* | *Periods Covered* |
|---|---|
| .10 | 19 |
| .155 | 12 |
| .18 | 10 |
| .20 | 9 |
| .30 | 3 |

In order to test the equation above using the information supplied by Figure 4–4, substitute the first set of numbers in the sequence, i.e., $A = .10$ and $N = 19$ periods.

$$A = \frac{2}{N + 1} = \frac{2}{19 + 1} = \frac{2}{20} = .10$$

$$N = \frac{2 - A}{A} = \frac{2 - .10}{.10} = \frac{1.90}{.10} = 19$$

or, where A = .18 and N = 10 periods

$$A = \frac{2}{N + 1} = \frac{2}{10 + 1} = \frac{2}{11} = .18$$

$$N = \frac{2 - A}{A} = \frac{2 - .18}{.18} = \frac{.182}{.18} = 10$$

Whether you wish to calculate the alpha factor or the number of periods included, neither presents any difficulty.

What is at issue here is certainly more important than mere mathematical exercises. You must ensure that whatever alpha factor you choose parallels your demand history. Trial and error would work, but you can't afford that much time. It is evident in Figure 4–4 that the lower the alpha number, the greater the number of moving-average periods represented. This is reasonable because the more time periods that are included, the more reliance you are placing on your basic forecast, not current-demand patterns.

The choice of the appropriate alpha factor is based on the relative stability of the demand cycle. In a mature operation, with a flat or recurring forecast, a low alpha factor is best because more emphasis is placed on the old forecast. When there are abrupt changes in demand—as with fashion cycles or fads—a high alpha factor is more appropriate.

Like the moving average, exponential smoothing is not without problems that must be addressed before adopting it as a forecast tool. First—like other calculations that are based on historical data—exponential smoothing forecasts items independently, which does not lend itself well to dependent-demand items. Second, choosing an incorrect alpha factor leads to incorrect forecasts and might cause damage to customer service. Third, the alpha factor, once chosen, tends to remain static and may not accurately reflect changes in business activity. Fourth, it cannot detect trends. And finally, it cannot detect seasonality. (Seasonality is a problem with *any* forecast, and its impact will probably have to be factored into any projections. This will be discussed later in the book.)

## USING FORECASTS

### Second-Order Smoothing

Back in the heyday of scientific inventory management, operations research people came up with a system of refining exponential smoothing to detect trends. This was called *second-*

*order smoothing*, which involved another step, called the beta factor. Without explaining the mechanics of the formula, suffice it to say that it has been pretty well discredited as a forecasting tool.

There are several reasons for this. First, and by no means least, it was extremely confusing—which, of course, was all the more reason for the consuming interest of theorists who believe that solutions to business problems can be quantified by the use of neat little formulas. The second reason is that it did not prove to be very practical in detecting the trends it was designed to detect, especially in the short run. Third, a better tool was discovered: the tracking signal.

## Tracking Signal

The tracking signal is a function of the *running sum of forecast error* (RSFE) divided by the *mean absolute deviation* (MAD). It is certainly not as complex as the terms suggest, but it does require attention to detail. Determining the tracking signal involves taking deviations from the forecast and treating them differently. Mean absolute deviation involves taking the average (mean) of the deviations from the forecast, both above and below the forecast and treating them in their absolute sense, which means disregarding the plus or minus signs. The total of the absolute deviations divided by the number of periods gives the average of the deviations per period.

To obtain the running sum of forecast error, one records each month's activity and adds to or subtracts from the actual or arithmetic deviation to the previous running sum to get a new sum. Obviously the RSFE must cover the same number of time periods as does the MAD, but the time periods themselves do not figure in the calculations. The intent is only to measure the total impact of the deviations. The tracking signal, therefore, is obtained by simple division.

$$\text{Tracking signal} = \frac{\text{RSFE}}{\text{MAD}}$$

The result indicates the direction of the trend. The higher the number, the greater the need for corrective action. Zero is the ideal result. Figure 4–5 examines a simulated forecast where the original forecast was 500. This was exponentially smoothed for 12 months, and you can see how the tracking signal reacted.

## Analyzing the Data

What does this rather routine forecast reveal? First, it shows that the exponentially smoothed forecast tends to lag demand. As pointed out earlier, this particular malady afflicts all time-series projections like the moving average. It may be better to simulate an alpha factor of .2 and a forecast weight of .8 and see if this shows a faster reaction time. It would, for example, forecast 553 for period 5 and, using that figure as the old forecast when moving forward, calculate 566 for period 6. These figures better reflect increasing sales activity.

There is not sufficient background information to develop a moving-average series for the model forecast. At the end of the first 12 months it computes to a simple average of 595,

**FIGURE 4–5**
**Yearly Forecast and Activity**

*a (sales) = .1* *forecast = .9*

| *Month* | *Forecast* | *Sales* | *Deviation* | *MAD* | *RSFE* |
|---|---|---|---|---|---|
| 1 | 500 | 464 | −36 | 36 | −36 |
| 2 | 496 | 330 | −166 | 101 | −202 |
| 3 | 479 | 474 | −5 | 69 | −207 |
| 4 | 479 | 847 | 368 | 146.25 | 161 |
| 5 | 516 | 618 | 102 | 135.4 | 263 |
| 6 | 526 | 772 | 246 | 153.8 | 509 |
| 7 | 551 | 573 | 22 | 135 | 531 |
| 8 | 553 | 432 | −121 | 133.25 | 410 |
| 9 | 541 | 938 | 397 | 162.56 | 807 |
| 10 | 580 | 642 | 62 | 152.25 | 869 |
| 11 | 586 | 750 | 164 | 153.55 | 1,033 |
| 12 | 602 | 294 | −308 | 166.42 | 725 |
| Total | 6,409 | 7,134 | 1997 | 166.42 | 725 |

$$\text{Tracking signal} = \frac{725}{166.42} = 4.36$$

Source: G. W. Plossl and O. W. Wight, *Production and Inventory Control: Principles and Techniques* (Englewood Cliffs, N.J.: Prentice Hall, 1967), p. 115.

which becomes the base number for month one as you move into another year.

Another thing that can be gleaned from the typical forecast in Figure 4–5 is that it provides the inputs to illustrate the tracking-signal calculations. The tracking signal (RSFE/MAD) is in the middle range. Most practitioners use 3–7 as an acceptable range, so 4.36 would probably not generate an exception report. But this depends upon the importance of the item, what the forecast review policy is, and, most important, human judgment. Should there have been a previous computation at the end of period 11, the tracking signal would have been much higher: 6.73. At this point the computer would probably have flagged this for review and possible corrective action, which would be to change the alpha factor to a figure greater than .1 and recalculate the MAD and RSFE to see what effect this would have on the tracking signal. Then one would update the forecast by adding 10 percent to the forecast, 10 percent being the difference between the annual forecast of 6,409 and the annual sales activity of 7,134.

Bear in mind that the foregoing is for illustration purposes only. As a practical matter the forecaster would not wait until the end of the year to find out how the forecast had performed, unless the information will be used as input data for the coming year's forecast, as would be the case in the first year of a moving-average forecast.

In any event, sufficient data must exist in order to make the tracking signal meaningful. It is assumed that Figure 4–5 does not cover the first year's forecast and activity and that historical data exist.

Finally, if the 12-month record of actual versus forecast is fairly typical of the annual demand cycle, it would appear that this product is seasonal. It will be shown shortly how seasonality alters the forecast in a positive direction.

## Demand Filter

If this is a seasonal product, attention is called to periods 8 and 9. These show a swing of over 500 units within the two periods, which suggests something may have happened during period

8 to reduce sales to an unusually low level, which then was made up in period 9, a month of abnormally high activity. To confirm this, a technique called the *demand filter* could be employed. This method calls for monitoring incoming orders, which may provide a more realistic picture of true demand than outgoing sales. Monitoring incoming sales in this manner can also call attention to large, nonrecurring orders that would tend to cause a lump in the sales figure for the month in which they are shipped, thus distorting future forecasts. The demand filter generally works best for stock items and is considered to be an alternative to the tracking signal. (It is always useful to use the demand filter when it is necessary to ensure incoming orders are falling within the normal range.)

## Seasonality

If there is a seasonal pattern of monthly sales volumes, it is mandatory to consider this when calculating forecasts. If records show sales of an item climbing from 8 percent to 10 percent, peaking at 15 percent above average, and then declining to bottom out at 4 percent, it is unreasonable to project a level forecast for all periods. The seasonality factor is an extremely useful and simple tool to reduce forecast error.

You begin by compiling a list of sales figures by month for a product or product class for the past 4–5 years to obtain a seasonal index. This may seem like an onerous task, but the information generated is as important as the information making up the basic forecast. Figure 4–6 shows a typical seasonal index.

You will note that the index is stated in percentages, not whole numbers. There are several reasons for this. First, the product line is unlikely to remain static for the entire five years. Additions, deletions, and changes have to be considered, whether the category is units or dollars. The same is true if the product line is built or assembled in or distributed from several geographic locations. Second, the percentages are important to the seasonality formula as one of the variables.

Using data for more than five years is clearly unnecessary and may even be confusing. If information is readily available beyond five years, you may wish to use it to make trial

**FIGURE 4–6**
**Seasonal Indexes**

| *Month* | *Y-1* | *Y-2* | *Y-3* | *Y-4* | *Y-5* | *Average %* |
|---|---|---|---|---|---|---|
| January | 7.48% | 5.46% | 6.36% | 6.56% | 6.54% | 6.50 |
| February | 8.24 | 4.70 | 6.68 | 7.35 | 5.78 | 6.55 |
| March | 8.27 | 7.30 | 8.84 | 7.90 | 8.03 | 8.05 |
| April | 7.95 | 9.34 | 9.40 | 8.79 | 9.29 | 8.95 |
| May | 9.66 | 9.65 | 10.50 | 11.11 | 10.44 | 10.30 |
| June | 10.00 | 11.30 | 10.60 | 8.79 | 9.71 | 10.10 |
| July | 9.06 | 8.70 | 7.66 | 9.01 | 9.96 | 8.90 |
| August | 9.85 | 11.10 | 10.80 | 10.12 | 11.00 | 10.60 |
| September | 9.71 | 9.50 | 8.15 | 9.70 | 7.90 | 9.00 |
| October | 6.65 | 8.72 | 8.09 | 8.78 | 8.95 | 8.25 |
| November | 7.80 | 8.08 | 7.64 | 7.34 | 7.72 | 7.70 |
| December | 5.12 | 6.15 | 5.13 | 4.54 | 4.64 | 5.10 |
| Total | 99.79 | 100.00 | 99.85 | 99.99 | 99.96 | 100.00 |

Source: G. W. Plossl and O. W. Wight, *Production and Inventory Control: Principles and Techniques* (Englewood Cliffs, N.J.: Prentice Hall, 1967), p. 34.

simulations to confirm your finding, but these should not become part of your working numbers. And data for less than three years may not be representative enough for the intended purpose. Variations in the level of sales are likewise unimportant. Of interest is only the interaction of one month on the other months being studied relative to an annual activity level. In this relationship, increasing or decreasing sales have no effect.

After the seasonal figures have been compiled, each month's activity is stated as a percentage of the annual sales total. This step is repeated for each period. The individual months are added together and divided by 5. This gives an overall average by month of the annual activity for the entire five years.

After these data have been input into a computer, the next step is to quantify the monthly time periods. The average month is 8.33 percent (100/12) of a year. Exponential smoothing has provided the basic forecast. This number multiplied by the overall monthly average and divided by the average monthly figure will yield the seasonally adjusted forecast.

For example, assume for a given month that by using exponential smoothing a monthly forecast of 6,000 units is

**FIGURE 4–7**
**Seasonally Adjusted Forecast**

| *Month* | *Forecast* | *Sales* | *Index* | *New Forecast* | *Deviation* |
|---|---|---|---|---|---|
| 1 | 500 | 464 | 6.50 | 391 | 73 |
| 2 | 496 | 330 | 4.63 | 276 | 54 |
| 3 | 479 | 474 | 6.64 | 382 | 92 |
| 4 | 479 | 847 | 11.87 | 683 | 164 |
| 5 | 516 | 618 | 8.66 | 537 | 81 |
| 6 | 526 | 772 | 10.82 | 684 | 88 |
| 7 | 551 | 573 | 8.03 | 532 | 41 |
| 8 | 553 | 432 | 6.06 | 403 | 29 |
| 9 | 541 | 938 | 13.15 | 855 | 83 |
| 10 | 580 | 642 | 9.00 | 627 | 15 |
| 11 | 586 | 750 | 10.51 | 740 | 10 |
| 12 | 602 | 294 | 4.13 | 299 | 5 |
| Total | 6,409 | 7,134 | 100.00 | 6,409 | 725 |

projected. The seasonal index for this month is 10.12 percent of the year's sales. The resulting equation would be

$$\frac{\text{Seasonal index}}{\text{Monthly average}} \times \frac{x}{\text{Forecast}} = \frac{10.12}{8.33} \times \frac{x}{6000}$$
$$x = 7{,}290$$

The monthly forecast adjusted for seasonality has been increased from 6,000 to a more natural 7,290.

For purpose of illustration, it is assumed that the numbers in Figure 4–5 are representative for five years. Figure 4–7 shows a year's activity with a seasonally adjusted forecast.

It is evident from Figures 4–5 and 4–7 that the old basic forecast totals and the new seasonally adjusted forecast totals are exactly the same for the 12 periods, as would be expected. They are merely functions of the base numbers allocated differently. Note also that the forecast understates the actual sales for 11 of the 12 periods. The effects of lagging notwithstanding, this still seems to indicate that the original forecast numbers were projected too low. Therefore, it would seem reasonable to forecast a higher level of sales. That is precisely what the RSFE would have suggested long before the former forecast. For just as the basic forecast and the seasonal forecast have the same total, so is the running sum of forecast error the same. Due to seasonality, the MAD of this new forecast is much lower (61). So the tracking signal would generate an exception report suggesting corrective action months sooner than would the basic forecast.

## Exponential Smoothing and Seasonality

Using seasonal indexes to forecast by the exponential-smoothing method requires a departure from the norms discussed so far. The rationale is that smoothing must be done using equivalent data in order to average correctly. All data generating a forecast adjusted by seasonal indexes must be adjusted against a common base (an average month)—in other words, it must be deseasonalized. Use as an example period 10 on the chart. The forecast is 580, and the sales for the period are 642. In order to calculate a forecast for period 11, the standard exponential-smoothing formula is employed:

| | | |
|---|---|---|
| Old forecast | $580 \times .9 =$ | 522 |
| Sales | $642 \times .1 =$ | 64.2 |
| New forecast | | 586.2, or 586 |

As can be seen in Figure 4–7, the monthly forecast for period 11 is indeed 586. Moreover, by using the seasonal adjustment, it can be seen that, historically, period 10 accounts for 9 percent (.09) of the total annual sales, which would have increased the forecast from 580 to 627. This also has occurred. What is now needed is to find out what the sales would have been if this were an average month. This can be done by reversing the seasonality formula:

$$\frac{\text{Seasonal adjustment}}{\text{Average month}} \times \frac{\text{Sales}}{x} = \frac{9.00}{8.33} \times \frac{642}{x} = 594.2$$

which indicates that the sales of 642 would be the equivalent of 594 in an average month with the seasonality netted out of the sales number. Interestingly enough, this is the simple average computed by dividing the annual sales by 12. This number, 594, replaces the 642 when calculating the new forecast for period 11:

| | | |
|---|---|---|
| Old forecast | $580 \times .9 =$ | 522 |
| Sales | $594 \times .1 =$ | 59.4 |
| New forecast | | 581.4 |

The new figure of 581.4 replaces 586.2 as the base forecast. To seasonally adjust period 11, whose index is 10.51, the standard seasonality formula is used:

$$\frac{\text{Seasonal adjustment}}{\text{Average month}} \times \frac{x}{\text{Forecast}} = \frac{10.51}{8.33} \times \frac{x}{581.4} = 734$$

This series of equations must be repeated for each period when generating a new forecast. It should be noted that 734 differs slightly from the forecast of 740 in Figure 4–7. Because the chart was calculated only with exponential smoothing, the average month was not netted out as would be the case in actual practice.

If the foregoing seems unduly complex and confusing, you have made a cogent observation. To repeat what was stated at the outset: The single most powerful and simple short-term forecasting tool is the moving average. If you want to seasonally adjust, you simply multiply the new average times the seasonal index to get the new month's forecast.

## Double Moving Average

Earlier the statement was made that despite the positive aspects of the moving average it is not without drawbacks. One of these drawbacks is that the moving average cannot effectively recognize and correct for trends, either up or down. Forecasting using historical data assumes that the future will be like the past, but what do you do if it isn't? To account for this, the technique to apply is the double moving average. Simply put, this is a single moving average of a single moving average. Put another way, sales are totaled for the periods that are included and divided by the number of periods, and then the deviations of actual versus forecast are totaled for each of the periods and again divided by the number of periods.

When numbers begin piling up on either side of the forecast, this calculated quantity can be added to or subtracted from the monthly forecast to arrive at a new trend-adjusted forecast. It can also be used as a flagging device to highlight demand changes on an exception report to be used for possible forecast revision. Referring back to Figure 4–7, we see that sales activity in 11 of the 12 periods is in excess of the forecast, which is certainly an indication that something is going on that requires human intervention.

It may seem that the double moving average is simply a running sum of forecast errors, and actually the two are quite similar, but the double moving average has several advantages. First, it is simpler to use. A continuing mean absolute deviation needs to be calculated after each period of demand. It may be a good idea to do this in any event, but it doesn't have to be done. Furthermore, the double moving average is distinctly more practicable. The tracking signal calculated by the RSFE/MAD provides only an indication that something is happening, while the double moving average allows you to do something about it. The netting-out process may show a trend in the making and allow you to correct for it.

Finally, the raw data for computing the double moving average can be used in an alternative manner. Because recent demand is more important when tracking trends, it may be possible to get a better forecast number using exponential smoothing. A high alpha factor would probably be chosen, especially if your gut feeling suggests the trend may continue.

### Moving Average and Seasonality

From a common sense point of view, using the moving average to depict seasonality beats any other technique by a wide margin. It is amazingly simple, although it does require a mass of data. But once the moving average has been computed, you will then have the average (base) month number. It does not have to be further deseasonalized in order to calculate the forecast. The process of averaging has already done this. This number times the seasonal-adjustment factor will produce the new monthly forecast.

It is apparent in the foregoing that the underlying objective is to get the maximum output from the least amount of data input. To be sure, the use of the moving average and all its corollaries requires the storage and manipulation of great amounts of data. This is always mentioned by its most persistent detractors, but it is no longer a problem. Practically every computer system in use today can handle the data with ease. Retrieval and processing time, however, are something else. (But, this has to be a management decision.) With increasing pressure from competition on the one hand and the need to get the biggest bang from your corporate buck on the other, it may take a little longer to do it right the first time. But if you feel you have a good handle on your forecasting problems and the processes available to help solve them, you may well have a leg up on everyone else.

## APPLICATIONS

Up to now, we have discussed in some detail the most commonly used forecasting techniques: the moving average, exponential smoothing, the demand filter, the tracking signal, and the seasonal index. Whether one wants to use one, two, or all of them is strictly a matter of choice. Obviously that choice will depend upon which system or systems allow for the least amount of forecasting error. Or you might choose to purchase a focus-forecasting package that includes the techniques described here as well as a host of others that are part of the software.

As in the economic-order quantity calculation, the forecast number generated is an independent one. Independent demand, a concept that will be more fully developed later, stands by itself, bearing no relationship to any other needed to make any product complete or functional. This is opposed to dependent demand, whose use depends upon the presence of other, coupling items. The independent status makes it a perfect application for forecasting finished goods, whether they are stocked as end items or in the finished component stage and are stored at the plant or in branch warehouses.

## Types of Manufacturing

The use of forecasting depends upon what type of business a firm is involved in. There are three general classes: a production shop, which makes parts to stock; a job shop, which makes them to order; and a third classification, which is called a job shop and is really a combination of the two. Over time a job shop may develop a variety of items it markets as stock items and also (as in the case of a sheet metal shop that does casework) fabricate modules from standard bills of material.

## Time-Phased Order Point

In the case of forecasting for finished goods, service parts, or other items with independent demand, the only realistic control is the time-phased order point system. Time-phasing here as well as time-phased material requirements denotes mainly planning for the future using time sequences. As a matter of fact, a subsystem under this heading has come to be known as *distribution requirements planning* (DRP). Figure 4–8 outlines the requirements of a distribution plan. The rudiments of this control stems from the timeworn order point formula: Order point equals demand during lead time plus safety stock. Or

$$OP = D_{lt} + SS$$

Lead time was once assumed to be fixed, or at least an independent variable. Now, in a time-phased mode, whether lead time is fixed or variable is disregarded—it is only required that it be known. In this context it becomes a function of priority planning and is updated regularly. More fundamentally, however, your aim is to control lead times instead of the other way around.

Because time-phased safety stock has a negative impact on inventory investment, it must be used only where specifically required and at a level based on recent history and common sense. Whereas safety stock was once viewed as a necessary evil that was somehow automatically built into the system, it is now the antithesis of modern inventory management. Despite this fact, there are still certain instances where safety stock is not only necessary but advisable:

**FIGURE 4–8**
**Distribution Requirements Planning**

| | | Period | | | | | | |
|---|---|---|---|---|---|---|---|---|
| | | *1* | *2* | *3* | *4* | *5* | *6* | *7* |
| Forecast requirements | | 20 | 20 | 20 | 20 | 30 | 30 | 30 |
| Scheduled receipts | | | 60 | | | | | |
| Projected available balance | 45 | 25 | 65 | 45 | 25 | 55 | 25 | 55 |
| Planned shipments | | | | 60 | | 60 | | |

Safety stock = 20, shipping quantity = 60, lead time = 2.

Source: T. E. Vollman, W. L. Berry, and D. C. Whybark, *Manufacturing Planning and Control Systems* (Homewood, Ill. Dow-Jones Irwin, 1984), p. 672.

1. The most obvious case is protection against variations in demand during lead time. The extent of forecast error mandates a given level.
2. Where lead time is long and fixed, where transit time is a substantial part of lead time (e.g., shipments from overseas or transported over very long distances such as rail).
3. At the master schedule level. Because of the uncertainties that exist before production—scrap, poor quality, etc.—it is usually advisable to produce something in excess of the forecast. *This is only true at the master schedule level.* Rarely—never, if possible—should safety stock be carried at the actual requirement level.

## SETTING SERVICE LEVELS

Central to any discussion of the time-phased order point is the attention given to the calculation of safety stock. For example, assume an order point of 600 units and a four-week lead time. With no safety stock, the law of averages says demand will be less than 600 half the time, more than 600 the other half. Obviously there is no concern with demand that is less than 600, so it can be stated that an order point of 600 units will produce a service level of 50 percent. The question now becomes: To increase the service level, how much stock must be added to the 600 units?

At this point it is important to know what the average deviation is when demand exceeds 600. Figure 4–5 shows that the mean absolute deviation (MAD) is 166, so the average demand in any month can be 600, plus-or-minus 166. If you add 166 units to the order point, you have added 1 MAD, producing an order point of 766. (One MAD is roughly equal to 1 standard deviation (Σ), but a MAD is much more convenient to work with.) The addition of 166 units raises the service level to 80 percent; 2 MADs provide a service level of about 95 percent.

$$50\% + 1\text{ MAD }(30\%) = 80\%$$

$$50\% + 1\text{ MAD }(30\%) + 1\text{ MAD }(15\%) = 95\%$$

It's easy to see that while more MADs raise the stock to very high levels, they would contribute very little to improved service. The normal curve with the MADs plotted on it is graphically presented in Figure 4–9.

The area to the left of the mean (average) demand line is the area that is below forecast, which is of no concern. It is the area to the right of the mean that figures in the calculation of safety stock because these occurrences exceed forecast.

As stated previously, the values of the standard deviation and the MAD are not exactly the same. Because the MAD is much simpler to use, a conversion table is necessary to change standard deviations to MADs before setting the level of desired service. This conversion number (safety factor) is then multiplied by the MAD to get the safety stock level.

That is why the process is begun by first calculating the MAD and then applying the desired service level.

The desired service level, or amount of safety stock carried, is calculated statistically from randomly occurring data. One might ask whether it works in actual practice. In order to test the validity of this concept, refer to Figure 4–7. One MAD (766) covers nearly 10 periods, while 80 percent of the year calculates out to 9.6 months.

What remains is to establish the service level desired—80 percent appears too low, but 95 percent carries too much ad-

**FIGURE 4–9**
**Normal Curve and Four MADs***

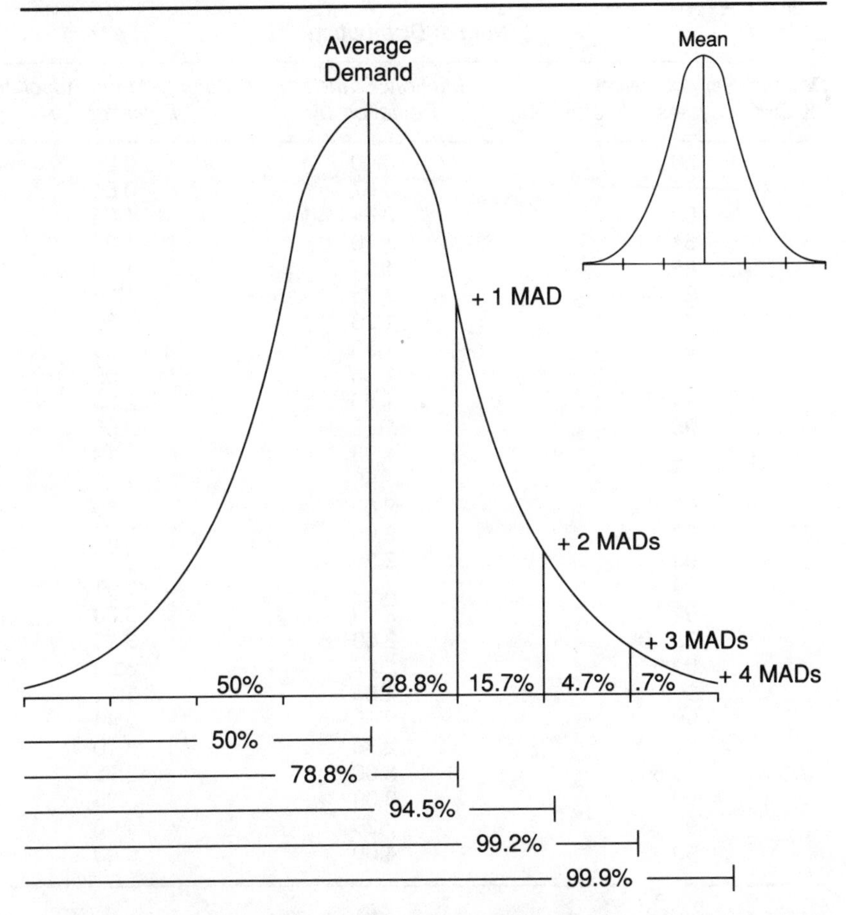


*There are approximately 1¼ MADs in each standard deviation (Σ), i.e., Σ = 1.25 MAD. This will show the real statistical properties of the normal curve or normal (distribution). One standard deviation from the mean encompasses 34 percent of random occurrences; and two standard deviations, an additional 14 percent, leaving only 2 percent outside this range (see diagram). These values are the same, either plus or minus, on either side of the center line.

ditional stock. All standard textbooks on this subject include charts fractionalizing MADs to given service levels (see Figure 4–10). If the desired service level is 90 percent, you would use 1.60 MADs to compute a new order point:

**FIGURE 4–10**
**Safety Stock Determination**

| *Normal Distribution* | | |
|---|---|---|
| *Desired Service Level (% Order cycles w/o stockout)* | *Multiply Standard Deviation by* | *Multiply Mean Absolute Deviation by* |
| 50. | 0.00 | 0.00 |
| 75. | 0.67 | 0.84 |
| 80. | 0.84 | 1.05 |
| 84.13 | 1.00 | 1.25 |
| 85. | 1.04 | 1.30 |
| 89.44 | 1.25 | 1.56 |
| 90. | 1.28 | 1.60 |
| 93.32 | 1.50 | 1.88 |
| 94. | 1.56 | 1.95 |
| 94.52 | 1.60 | 2.00 |
| 95. | 1.65 | 2.06 |
| 96. | 1.75 | 2.19 |
| 97. | 1.88 | 2.35 |
| 97.72 | 2.00 | 2.50 |
| 98. | 2.05 | 2.56 |
| 98.61 | 2.20 | 2.75 |
| 99. | 2.33 | 2.91 |
| 99.18 | 2.40 | 3.00 |
| 99.38 | 2.50 | 3.13 |
| 99.5 | 2.57 | 3.20 |
| 99.6 | 2.65 | 3.31 |
| 99.7 | 2.75 | 3.44 |
| 99.8 | 2.88 | 3.60 |
| 99.86 | 3.00 | 3.75 |
| 99.9 | 3.09 | 3.85 |
| 99.93 | 3.20 | 4.00 |
| 99.99 | 4.00 | 5.00 |

Source: APICS Master Planning Certification Review Course V-63

$$166 \times 1.60 \text{ (MAD factor)} = 266$$

$$600 + 266 = 866 \text{ (new order point.)}$$

Referring again to Figure 4–7, the new order point of 866 will cover 11 of the 12 periods (91.6 percent).

Choosing a given level of service, therefore, mandates carrying a specific amount of safety stock. As a result, this has to be a management decision. Management must balance in some way the penalty of being out of stock with the cost of carrying extra insurance in the form of safety stock. As can be seen on the preceding page, the safety stock calculation is strictly a function of the mean absolute deviation. A demand history that is relatively flat and not given to wide swings month to month can be accommodated by a lower level of safety stock. However, as deviations increase, the increase in the amount of safety stock relative to a given service level may well tend to put undue strain on the company's resources.

This is yet another instance when common sense will transcend mathematical manipulations. Furthermore, because the figures used in the computations work out so closely with those in the sample forecasts, it must not be assumed that this is always so. The concept of stockouts and service levels is a statistical one and, as such, rarely work out so conveniently.

## Forecasting and Dependent Demand

The final consideration under forecast applications has been touched on previously, but is reiterated here for emphasis. It concerns using the forecast for dependent-demand items, which is never a good idea and should be used only as a last resort. The independent status of a forecast makes using it for ordering dependent items very risky. When items are produced for stock, the forecast becomes the plan and time-phasing net requirements from the master schedule is routine.

In a job shop environment it is the backlog that becomes the scheduling tool. If your backlog does not extend far enough out to encompass certain long-lead-time components, a forecast of usage is mandatory to provide for their being on hand when needed. For components with much commonality it may be convenient to forecast annual usage and place blanket orders for yearly quantity with periodic withdrawals, rather than placing many small orders for each requirement. There may be other circumstances that require special attention—e.g., threat of a work stoppage, ordering from a new supplier, etc.—and make using a forecast for ordering good sense. These will require a forecast because nothing better is available.

The same reasoning applies to safety stock. It should viewed as a concomitant to the forecast quantity, not as an expensive safety valve.

**Choosing the Forecast Tool**

Experience will probably dictate which, if any, forecasting model a user will adopt. Most companies using exponential smoothing will set their alpha factor at a rate that will reflect the same history as does the moving average. This is open to question, however. Even though a computer can handle these numbers with ease, it does nothing to diminish the confusion that results from the use of so many different equations to build a forecast. Simplicity is still the name of the game. Management will find it difficult to adopt a system they do not understand, and the moving average is so understandable.

Take the tracking signal as an example of non-simplicity. The equation using the RSFE over the MAD will do the job of tracking a forecast generated by exponential smoothing. The moving average has a built-in tracking signal requiring no additional computation. They both will lag behind a trend, so by the time the tracking signal recognizes the trend, it may well be past. In addition, many companies begin the year with a marketing forecast as the starting quantity and then smooth demand from month to month. If a correction needs to be made, it may be four to six months before the computer generates a signal. On the other hand, the moving average will continually and automatically correct with no other signal being necessary.

The seasonal index is another method exponential smoothing seems to complicate. Continually returning to the old forecast, netting out seasonality to calculate an average month, and using these to compute a new month's forecast certainly runs a poor second to using the moving average times the monthly index formula.

Perhaps it's true that for the moving-average technique to be valid, demand must be relatively stable and deviations not too great, but, as stated earlier, there is no better forecasting system yet devised. The moving average coupled with good judgment will do the job.

## Forecasting Responsibility

Forecasting is an onerous task—boring and largely unfulfilling. At the same time, it is also one of the most basic tools in any production and inventory control system. In a larger sense, the building of a forecast is the responsibility of the Marketing Department because this will become a basic ingredient of the business plan. Still, it is intricately linked to the historical data that can only be provided by the material planning and control functions. Because forecasts are part of management's larger planning processes, they are better left to those individuals who possess the most complete knowledge of the marketplace. All too often, companies never clearly identify the purposes forecasts are intended to serve, in which case they do not get the support of top management that they require. It must be understood: Forecasts are a basic business requirement, and, whether they are formally or informally developed, a business cannot help but utilize them in some manner.

## SUMMARY

The foregoing discussion of forecasting techniques should not be construed as all-encompassing. There are several other formulas and models that might have been mentioned—e.g., the Box-Jenkins approach and triple smoothing—but they are too narrow in scope to be of great practical use. The purpose here was to set out in detail and compare the more useful techniques in actual application.

# CHAPTER 5

# THE MASTER PRODUCTION SCHEDULE

The master production schedule (MPS) in conjunction with material requirements planning (MRP) form the "drive train" of the manufacturing machine. The MPS itself is the preeminent module of all production and inventory control documentation. Regardless of what it is called and whether it is formal or informal, an MPS will be used by every firm engaged in manufacturing.

Many texts dealing with this subject start at the top of the master-planning hierarchy and examine in some detail the overall business plan. Then they proceed into production planning and from there into development of the MPS. Admittedly, as far as total planning is concerned, these are extremely important, but they will not be of concern here. The concern here is the development of the master production schedule as the start of the manufacturing cycle. To be sure, the MPS must be consistent with a higher scheme of things, but how management gauges its progress as the schedule rolls through time is primarily a matter of choice. Figure 5–2 at the end of this chapter outlines the steps required in developing and using the MPS.

If MPS is the basic planning tool, it must conform to a set of conditions. First is a definition: The master production schedule is a statement of the anticipated build schedule constrained by time periods—in other words, it details what is expected to be produced and when. In short, it is equal to the total current demand upon a production facility.

The MPS is not the same as a forecast—although sometimes they will be identical—because they each perform distinct functions. First, the time intervals covered are not the same: The MPS generally spans a week, while the forecast is for a longer interval. Second, the forecast is only an estimate of demand, and the MPS is the actual anticipated plan for production. Finally, the forecast does not involve any awareness of capacity constraints, while the MPS becomes committed through their identification.

The MPS must be stated in terms of units and the production rate in units per hour, day, week, etc. Higher-order business plans are usually stated in dollars. This is the first time specific units are identified. Dollar volume would carry no significance. The units scheduled are always end item products. Stated another way, they are zero-level items on the bill of material. It doesn't matter whether these are finished goods, major assemblies, or component-level repair or service parts as long as they are end items identifiable for their intended purposes and in the quantity specified.

## MASTER PRODUCTION SCHEDULE INPUTS

Simply stated, there is only one input into MPS: demand—forecasted demand and actual demand. The latter, obviously, is of greater concern. It includes orders from any source: customer orders for finished goods and service parts, interplant orders, replenishment orders from branch warehouses—anything that goes out the door that results in an invoice. The holes, so to speak, can then be filled in from the forecast. For example, if the forecast is 500 and "booked" orders are 300 in a given period, the master schedule will show 500. On the other hand, if orders total 700 against the same 500 forecast, an attempt will be made to accommodate this amount, assuming material and capacity availability. If not, the most reasonable alternative is to shove something into a later time period. More on this later. Figure 5–1 lists all the contributors to demand.

An important point should be stressed here. This is the only place in the timing of things where there will be some latitude in the customer's requested delivery date. If an order

**FIGURE 5–1**

| *Sources of Demand* |
|---|
| Customer orders |
| Dealer orders |
| Master scheduled-item forecast |
| Safety stock |
| Interplant requirements and orders |
| Service and repair requirements; samples |
| Orders for stock |
| Pipeline fill |
| Finished-goods warehouse requirements |
| Branch warehouse requirements |

carries a ship date of, say, the 20th of the month, it can be established immediately if this is a valid date. If it is not, an alternate can be negotiated. Moreover, if the ship date presents a problem, Marketing must be advised immediately, not four days after it was supposed to ship.

Some writers contend that the production plan should provide input into the MPS and that the summation of the entire schedule be a reflection of this plan. This contention can be somewhat misleading. If the entire MPS is a breakdown of the forecast, this will happen because the plan and the forecast were generated using similar data. But the random pattern of incoming orders does not allow this luxury. So when a schedule takes shape and is not consistent with the production plan, it is not tolerable to arbitrarily alter the MPS in order to force it into agreement with the production plan. Once again, consistent with customer needs, the rule of common sense always takes precedence.

## MASTER PRODUCTION SCHEDULE OUTPUTS

### Material Requirements Planning

The master production schedule is the driving force of MRP, although there are several other minor inputs that will be discussed later.

### Rough-Cut Capacity Planning

As the completed MPS passes into MRP for the requirements explosion and subsequent order issuing and capacity investigation, any shortfall can result in a change in schedule. To avoid duplication of effort, the MPS is first tested against critical work stations or bottleneck situations. The capacity required in the MPS is checked against the capacity available at the work centers by the use of the product load profile to see if any serious overload situation can be detected. (The construction of a product load profile is explained in detail under Capacity Planning.) The product load profile imparts a simulated quantity of one and this, times the quantity on the order, computes the hours required to process a given order onto whichever work station the order is assigned. Other, possibly alternative, work centers are similarly quantified to see if an underload exists. When all these data are assembled, simulation is carried out to find the most efficient and level MPS. This, then, is sent to the material planners for them to work their magic on.

### Final-Assembly Schedule

Whenever a process is required to assemble a group of parts according to the customer's order, a final-assembly schedule (FAS) is required to accomplish this in a logical sequence. An FAS is most often used in make-to-order or assemble-to-order situations. Characteristically, the FAS covers a shorter time span than the MPS does—usually a day, never more than several days. Because the FAS is merely an extension of the MPS, it necessarily has the same importance.

## TYPES OF MANUFACTURING

The construction of the master production schedule is determined by the manufacturing environment as well as by the timing involved and the product structure. The manufacturing environment may be one or a combination of the following three types:

1. Repetitive manufacturing, where the same products are made on a continuing basis for a projected period of time. The number of units may vary, but the products themselves do not. Many consumer goods fall into this category. The MPS is created from a forecast with a good measure of executive opinion.
2. In a true job shop, also known as *discontinuous manufacturing,* products are made strictly to customers' specifications. In this category are specific types of machinery or tooling that may or may not be run again—automotive components under contract being a good example. The determining factor is who develops the prints. The MPS includes all in-house orders as of a given date and operates almost exclusively on an order backlog basis.
3. In a modified production shop, a line of standard products is manufactured to customer order. This is probably the most prevalent type of manufacturing today. Often a firm holds a wide variety of standard components at the semifinished or subassembly stage and assembles them into completed units as orders arrive. Or a company with its own line of products may wish to take on private labeling or other forms of subcontract work to use up excess capacity and increase efficiency. This would include items produced on customer-supplied tooling and equipment. The MPS is based on order backlog because, theoretically, nothing is made for stock. However, planning could be enhanced by some sort of forecast.

## MPS APPLICATIONS

When considering product demand, you must deal with some form of identification for each product. You must assign to each product some form of coding that will become its part number. This part number will be the same as or will directly access the end item, or zero-level part number, on its own bill of material, which is the medium for setting the entire manufacturing cycle into operation. This is true not only in the planning stage but throughout a product's life on the shop floor. This procedure is absolutely necessary; to do otherwise would cause crucial errors by obscuring a part's traceability. This is common sense in its simplest form.

The graveyard of failed MRP systems is replete with the bones of master production schedules in which the priority planning and control features were incomplete, either lacking in reality or hyperambitious. The former can be shown by means of a simple example. Portions of the weekly MPS are as follows:

Firm schedule – in days

| 1 | 2 | 3 | 4 | 5 |
|---|---|---|---|---|
| 100 | 100 | 100 | 100 | 100 |

Operation 1

Planned schedule – in days

| 1 | 2 | 3 | 4 | 5 |
|---|---|---|---|---|
| 100 | 100 | 100 | 100 | 100 |

Operation 2

Here are two successive weekly production schedules. The particular item and quantity are inconsequential, and it is assumed that material is on hand and capacity is available.

Early in period 1 (Monday) of the first week, Scheduling is notified that an important customer needs an order one week earlier than was originally promised. Because this is not an unusual request, Scheduling, after investigating the impact on the total schedule (and upon instructions from Marketing), reschedules the production of this part from period 5 (Friday) to period 2 (Tuesday) for the first operation:

Firm schedule – in days

| 1 | 2 | 3 | 4 | 5 |
|---|---|---|---|---|
| 100 | 100 | 100 | 100 | |
| | 100 | | | |

This leaves period 5 open to run something else. The next order in sequence is moved into this open slot.

In the following week (operation 2) this 100 is again moved forward from Friday on Tuesday, leaving Friday's slot free to run something else. The total demand for the two-week period would look like this:

Firm schedule – in days

| 1 | 2 | 3 | 4 | 5 |
|---|---|---|---|---|
| 100 | 100<br>100 | 100 | 100 | 100 |

Operation 1

Planned schedule – in days

| 1 | 2 | 3 | 4 | 5 |
|---|---|---|---|---|
| 100 | 100<br>100 | 100 | 100 | 100 |

Operation 2

The net effect has been to gain two days by compressing 12 days of demand into 10 days' capacity. Obviously you could increase capacity in the short run by scheduling overtime, but the effects of this rarely show up until some time in the future.

At a glance, this example seems absurd. A scheduler should not allow this to happen in the course of planning. However, as everyone knows from experience, it does happen. Although it may seem like an isolated and extreme example, because of the complexities of the real world of manufacturing and the myriad other responsibilities that comprise the workday, it is not difficult to see how this could occur.

Furthermore, one can rationalize that the system will eventually catch up with the situation. This is true, but it will probably be too late for corrective action and precipitated by a crisis. This situation is more likely to occur with manual systems, where continual adjustments of priorities to keep them valid is too burdensome. But in a correctly working computerized MRP this will not happen. The principle is elementary: If you put something in ahead of schedule, an equal amount must come out; if you reschedule period 5 into period 2, you must take the original quantity for period 2 and put it somewhere else—it is a simple tradeoff.

The foregoing is not the only way to create chaos out of order. There are plenty of others. You may intentionally overload the schedule so that the shop will have to work harder in order to meet it. This is self-defeating and a path to disaster. Or you may schedule Saturday production in hopes of catching up. This is less than ideal for two reasons. First, you are paying time-and-a-half to a morose work force whose minds are somewhere else. And second, half of them won't show up for work on Monday.

Due to their insidious nature, these problems generally are not recognized until they become all too obvious. The typical way to work through these problems is to hand them to a team of expeditors to sort out. But there is a lesson that is taught but rarely learned. If you are using expeditors in the shop, your system is not working. Common sense tells us that you're better off doing it right the first time.

## OTHER ASPECTS OF THE MPS

The only certainty one can have about a master production schedule is that in spite of the tediousness of constructing a complete schedule, it is bound to require changes. The Greek philosopher who said nothing is so constant as change must have worked in a manufacturing shop! Approaching the MPS with this in mind should help reduce the irritation caused by frequent changes in the schedule. Remember: This is the way the game is played.

One final comment is in order. Some practitioners have suggested that the master schedule should be built using EOQs. In the first place, to use the EOQ formula for anything practical is in itself a ridiculous notion. Then, to calculate the quantity of finished goods using the standard formula would sorely test the talents of a Merlin. Finally, to take a product comprised of many components, made in-house and/or purchased outside, with widely varying setup and ordering costs, unit costs that go all over the countryside, and differing annual usages, and then to think that you can come up with something meaningful is almost ludicrous.

## SUMMARY

The master production schedule is an organization's primary planning tool. The goal of customer service cannot be accomplished without proper management of this most important step in the manufacturing process. Nothing can begin without it.

Therefore, the MPS must be ambitious, yet attainable—the key word being *attainable.* The impulse to overload the MPS may sometimes be overwhelming, but it must be resisted, for the result will be chaos on the shop floor. People will not work harder in an attempt to reach unrealistic goals. They will resort to their own, informal priorities. In fact, as has been learned by applying the principle of Just-In-Time, a somewhat underloaded schedule may be a viable alternative because the objective is to look at throughput instead of efficiency.

Feedback, while not specifically mentioned, is implied throughout the entire section. Feedback is the scorecard, the only way to accurately measure performance, so rules must be established to maintain its timeliness. Variations of actual versus scheduled output must be processed to assess their impact on customer service. Feedback from requirements planning is likewise significant. Any impending problem regarding critical material could adversely affect the priority sequence of the MPS.

Keeping priorities valid is probably the thorniest responsibility a scheduler has to cope with. There always seems to be more necessary input at any given time than there is capacity to handle it. The inclination to pad a little extra into the MPS to "see if we can do it" must be avoided. The master production schedule is not a wish list; it is a detailed document that states not only what needs to be produced but also what is expected to be produced. The MPS is a reflection of the promises made to the customer of what can and will be done. When assembling the orders that will make up the schedule, this rule is inviolate.

**FIGURE 5–2**
**Master Production Schedule Schematic**

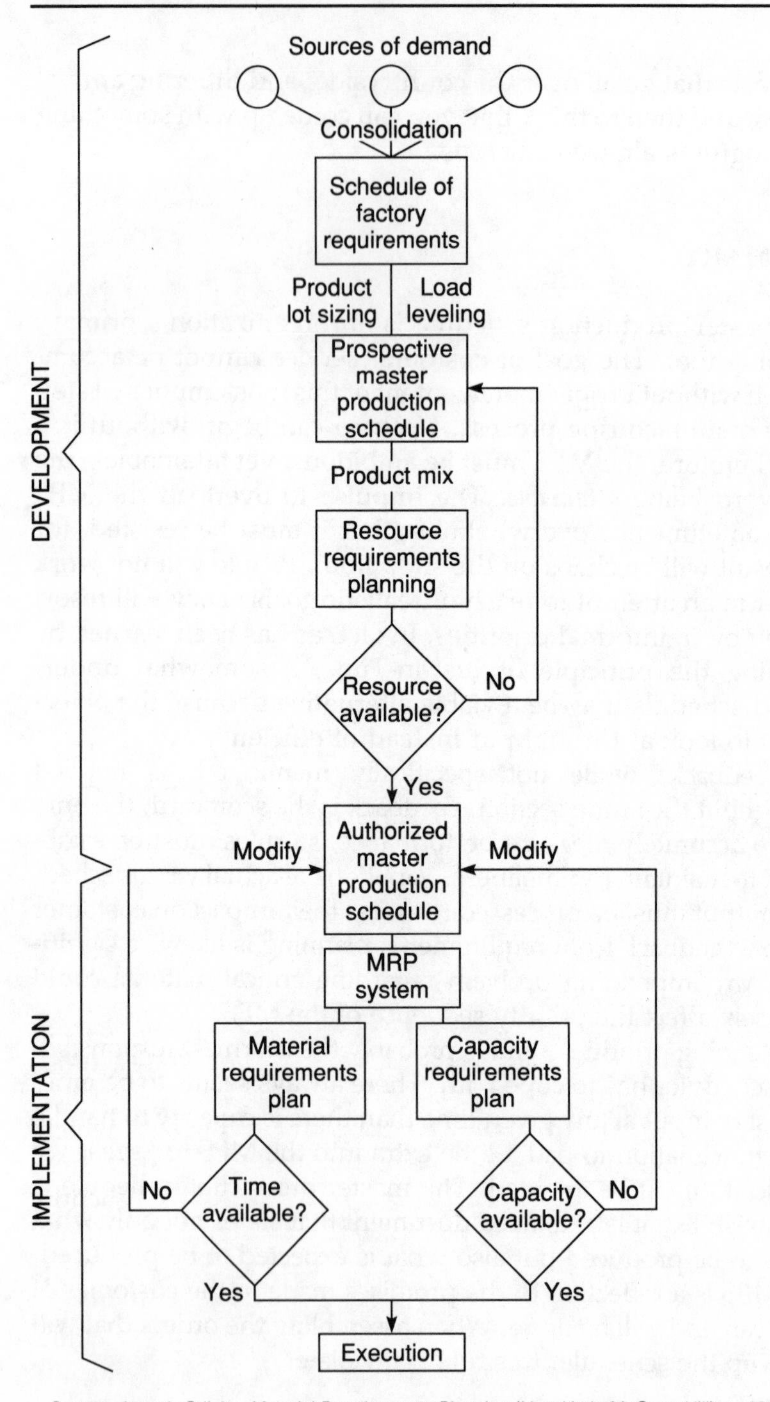


Source: Joseph Orlicky, *Material Requirements Planning* (New York: McGraw-Hill), p. 251.

# CHAPTER 6

# MATERIAL REQUIREMENTS PLANNING

When this writer was a graduate student, back in the mid-Jurassic period, he was told in a management class that business of any sort whatsoever had to fulfill three distinct functions and that, furthermore, no business could exist without them. The class was challenged that if anyone could name an enterprise that could survive without all three functions, the person would be awarded an automatic A.

The professor was, of course, referring to the

Production function: making a product or providing a service.

Sales function: exchanging and distributing the goods or services.

Financial function: funding the ongoing activities of the organization.

If this is held to be true (as it must be, because no one ever got the A), it is equally accurate to say that a manufacturer cannot exist without the function of inventory. Because it has value, manufacturing inventory is unique in the way in which it is handled, which is why many volumes have been written about the means to control it. So, just as inventory is a function of manufacturing, Material Requirements Planning is a function of inventory.

In our previous discussion of scientific inventory management, it was pointed out that the order point technique, while logical, proves unworkable because it lacks one very significant input. When the order point is reached and an order placed, there is no way of telling when, if ever, the part will be required again. Nor does it tell the scheduler whether the quantity ordered is too much or too little to handle future requirements. MRP seems to provide an answer to both puzzles simultaneously with a technique called *time-phasing*.

Time-phasing not only answers the questions of how much and when to order better than does the order point, it also determines the date when the requirements are needed. Basically time-phasing says that if there are no foreseeable requirements for a part in the production schedule, why order it and unnecessarily tie up capital, regardless of what the order point indicates? Then, when the schedule shows a requirement, an order is placed within the limits of lead time so that it will arrive just when it is needed. (This is termed *time-phasing* it.) Then, the order is placed for just the amount the schedule calls for (or maybe the period order quantity), regardless of what the artificially contrived EOQ calls for. Thus, if the requirement shows 300 and there is an EOQ for 550 (or worse, 250), the order is placed for 300 because that makes sense. A way of tracking time-phased requirements is shown in Figure 6–1.

As this figure plainly shows, nothing happens until period 17. The time-phased feature indicates that nothing need be done until period 15, when the order covering this requirement is released (assuming a two-week lead time). Under an MRP system, the order point is 0; the order quantity 50.

**FIGURE 6–1**
**Time-Phasing**

| | Week | | | | |
|---|---|---|---|---|---|
| | 14 | 15 | 16 | 17 | 18 |
| Gross Requirements | | | | 50 | |
| Scheduled Receipts | | | | | |
| Projected on Hand | | | | | |
| Planned Order Release | | 50 | | | |

## MATERIAL REQUIREMENTS PLANNING SYSTEM

In its earlier days MRP was thought to be just a superior inventory ordering technique. It answered the second half of the needs equation. Over time its meaning has expanded to represent a complete manufacturing scheduling and control system. This is the familiar closed-loop MRP (MRP II), which, ideally, encompasses most or all of the operational functions, linked together to achieve a more effective planning of the firm's resources. A graphical presentation of a formal closed-looped MRP (MRP II) system is shown at the end of this chapter in Figure 6–22.

The purpose of this chapter is not to attempt to construct a closed-loop, or MRP II, system, which follows logically after achieving common sense manufacturing. As the title of the chapter suggests, we will discuss materials management: the flow of goods from the source to the customer and the tools to deal with them effectively along the way. Besides, a successful closed-loop system evolves in progressive steps over a long period of time; it is not imposed on the troops from above.

Simply stated, material requirements planning (MRP) is the calculation of requirements in terms of both quantity and time that are needed to make the end items as specified by the master production schedule (MPS). As stated previously, some form of the MPS must be undertaken so that the MRP has continuity.

In order to understand the MRP process, it is necessary to examine the inputs and outputs of the system:

| *MRP Inputs* | *MRP Outputs* |
|---|---|
| Bill-of-material | Orders |
| Inventory on hand | |
| Components on order | |
| Planning data | |
| Master production schedule | |

You will note that conspicuous by its absence is forecasting, which has no place in MRP. It is only used when there is nothing better available. With MRP one can always calculate requirements exactly; no estimates are necessary. It is true that there are some narrowly defined situations where some forecasting might be of assistance, but you will recognize these and make allowances for them. Otherwise, forecasting is reserved for the estimation of unknown and independent demand.

## INDEPENDENT/DEPENDENT DEMAND

Essential to MRP is the ability to isolate independent demand from dependent demand, which is not at all difficult. An independent-demand item is one whose demand is not directly related to the demand for anything else. For the purposes of this discussion, these will be limited to specific, definable master-scheduled items, from finished goods to spare or service parts.

A dependent demand item, on the other hand, is one whose usage can be calculated from the requirement to build a higher-level item. It is a component of something else. A ball that goes into a bearing has dependent demand, as does the housing for an electric motor. One *might* think that the parent item, which is dependent upon the availability of lower-level components, should be referred to as having dependent demand, but this is not true. The status of an item is dependent because its usage depends on its availability (as well as the availability of other component parts) for making up a parent assembly, as shown below:

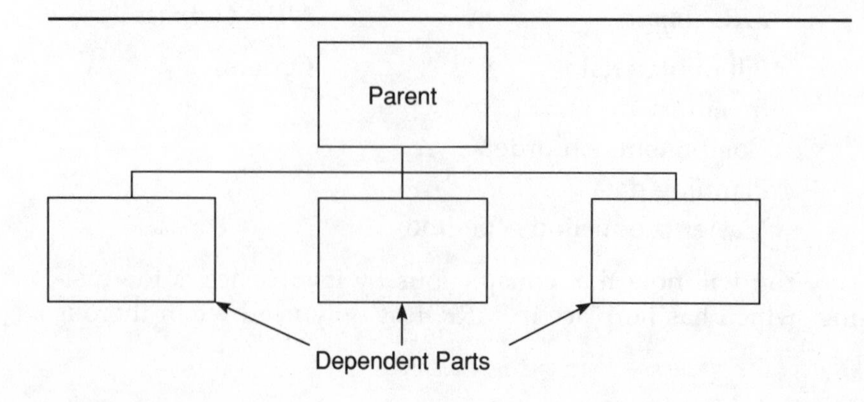

## INPUTS

### Bills of Material

From its designation on the MPS, the customer order or quantity in the forecast states the specific bill of material that will be called out to build a particular end item. This bill is a listing, by part number, of the individual piece parts that go into making one of the completed product. In other words, it is an ingredients list. This bill will also show the quantity per unit required for each parent part, how many parent parts are required to make up a still higher level assembly (a grandparent?), and so on to the zero-level end item. From these data, MRP can calculate the gross requirements to produce the scheduled quantity. An accurate and up-to-date product structure depicted in this fashion readily shows that any part on a level below the end item has dependent demand status somewhere within the production cycle.

### Part Numbering

The bill of material and the assignment of part numbers are the domain of the Engineering Department. And part numbering is the first and most important step in constructing a bill of material. Each part must have its own unique identifying digits, so a system of assigning numbers to parts must be decided upon. Part numbers can be entirely *nonsignificant*—that is, sequentially assigned (e.g., 1001, 1002, 1003), with no other meaning—or contain (in addition to the nonsignificant digits) a *significant* portion, usually a prefix of one or two digits, that relates the part to a class of parts (e.g., 10-1001, 10-1002, 10-1003, where the prefix, 10, identifies the parts as purchased, but no hyphen would appear). It is best to avoid letters when assigning part numbers. Practically everyone can identify all their parts with six or seven digits, which are less cumbersome to use, while still leaving plenty of room for expansion. Many computer systems once used a final check digit to confirm that the part number was correct, but today bar coding makes this unnecessary.

The importance of part numbering cannot be overemphasized. It is the only logical way to identify a large array of individual parts, and it forms the basis for all inventory transactions.

### Bill of Material Structuring

Product structuring and bill of material formatting are important for both ease in identification of relationships and for retrieval purposes. The bill should indicate how the product is put together. It need not be a blueprint for construction, but it should show the relationship of the various components as the product is being built. If one were to diagram a flow chart for the product, it would depict the zero (end) item at the top, with an increasing number of parts shown as the levels descend, thus producing the familiar "Christmas tree" configuration. The easiest and most common way to display this for planning purposes is through the indented bill of material (see Figure 6–2).

Obviously the indented bill of material is more common where the product is built in a series of levels linked together. Each component part is indented directly beneath its parent, with the quantity required per parent assembly just to the right. In this way the entire structure or any part of it can be displayed for observation and planning.

It is vitally important that the retrieval system via the bill of material explosion process described above is accompanied by a "where-used" capability. This where-used, or implosion, capability does precisely what the term implies: It indicates all the various process implications where the part is used, so that when there is a change in cost or an engineering change, the extent and effects of the change are readily apparent.

## BILL OF MATERIAL TYPES

### Phantom Bill

This is known by various names including *transient* or *blow-through bill.* It is so named because the parts or assemblies involved, although they do physically exist, are not recorded as

**FIGURE 6–2**
**Indented Bill of Materials**

BILL OF MATERIALS

PRODUCT 36401 LOCKER BASE 4162

PAGE 1 OF 1

| PART NO. | DESCRIPTION | NO. REQ'D. M | NO. REQ'D. P |
|---|---|---|---|
| 36401 | LOCKER BASE | | |
| 36001 | BASE - FINAL | | |
| 20010 | TOP | | |
| 10112 | 22 GA. 13 3/4 X 23 1/8 | 1 | |
| 20012 | BASE | | |
| 10112 | 22 GA. 13 3/4 X 23 2/8 | 1 | |
| 20020 | BACK | | |
| 10113 | 24 GA 11 13/16 X 31 | 1 | |
| 20043 | RIGHT SIDE | | |
| 10115 | 20 GA. 23 31/32 X 32 | 1 | |
| 20106 | SHELF | | |
| 10118 | 22 GA 13 3/16 X 21 | 1 | |
| 36107 | LEFT SIDE ASSEMBLY | | |
| 20068 | LEFT SIDE | | |
| 10115 | 20 GA. 23 31/32 X 32 | 1 | |
| 20005 | PADLOCK TONGUE | | |
| 10102 | 14 GA. 7/8 X 3 25/32 | 2 | |
| 35119 | DOOR ASSEMBLY | | |
| 62013 | #221 HINGE (TASSELL) | | 4 |
| 20223 | DOOR | | |
| 10120 | 18 GA. 12 X 16 | 2 | |
| 20004 | FINGER PULL | | |
| 10103 | 16 GA. 7/8 X 1 7/8 | 2 | |
| 62114 | POP RIVET SD 64 BS | | 4 |
| 62100 | BOLT PACKAGE | | |
| 62321 | BOLT 1/4-20 X 1/2 HEX HEAD | | 4 |
| 62068 | LAG BOLT 1/4 X 1 1/4 | | 4 |
| 85101 | CARTON | | 1 |

ASI ANGLE STEEL INC.
PLAINWELL, MIC
DATE:

KAL BLUE

inventory transactions into or out of stock. These parts would be common to a given category of end items or assemblies, but they would be made on the line on a one-to-one basis, with no lead time considered. Posting of these parts to stock records would be unnecessary.

Another application of the phantom bill would be where an additional operation, like painting or plating, is required following completion of the assembly. Although the painted part would carry a different part number, posting it into and out of stock would serve no good purpose.

For an example of this, refer to the indented bill of material (B/M) in Figure 6–2 for the 36401 Locker Base. The last assembly item is Bolt Package 62100. Although these bolt packages are actually assembled in the stockroom, they are delivered to final assembly in accordance with the quantity on the MPS. They are never posted to stock records and are not normally carried in inventory. The assembly B/M is kept there for their use only, and inventory is controlled using the two-bin method.

It should be noted, however, that although phantom assemblies are not posted to inventory records, lower-level parts that make them up are posted. They are planned for under the MRP procedure just as any other requirement would be. As you can see from the example, the two-bin method is merely an alternative way of using MRP.

## Planning Bills of Material

The planning bill of material serves a very useful purpose. In a repetitive environment, where product configuration may include different features based on a choice of available options, it would be impossible to forecast with any degree of accuracy the exact makeup of customer orders. In order to overcome this deficiency, a master scheduler can schedule through the shop all possible combinations of components or major assemblies that could conceivably go into a finished product. This is done using percentages (or decimal fraction amounts) that represent averages of the most recent demand as they relate to an end product. These quantities are what are master-scheduled.

**FIGURE 6–3**
**Planning Bill of Material**

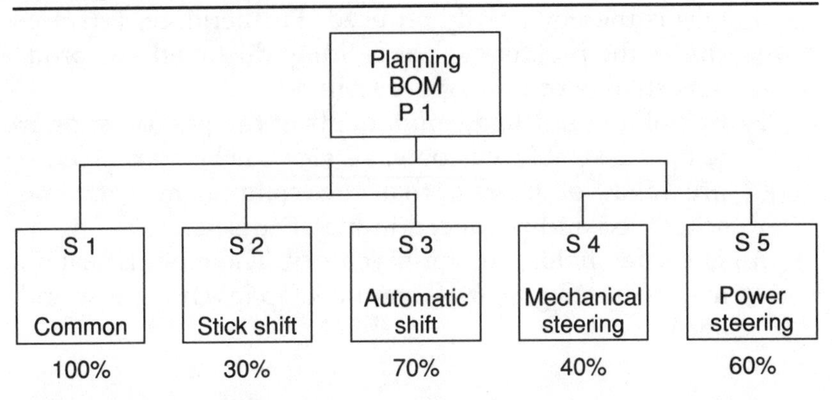


Source: APICS Material Requirements Planning Review Course

The planning bill, although it is a true bill of material, is actually a master scheduling device. Obviously it does not define a buildable product, but it is used to forecast. As such, lower-level components of the parent item are not usually shown because MRP would take over at this stage. Figure 6–3 shows a very simplified version of the planning bill of material.

## Modular Bills of Material

Modular bills are similar to planning bills in that they do not represent a specific end product. Instead, major components, or modules, are brought to the point just prior to final assembly, where the product can be built to whatever configuration called for on the customer order. According to the late Joe Orlicky, these modules are "promoted" to end item status.[1]

Modular bills of material perform three definite functions:

1. Constructing separate bills for each possible end item configuration would be unmanageable, especially if

---

[1]Joseph Orlicky, *Material Requirements Planning* (New York: McGraw-Hill), p. 224.

there were a number of options to choose from. By master-scheduling the modules only, the number of bills is thereby greatly reduced. Furthermore, this precludes the proliferation of building documents as products change or new options are added.

2. By holding finished components at the point just prior to final assembly, customer service will be maximized—provided, of course, that the component parts are scheduled and produced in matched sets.
3. The same holds true for inventory: The modularization feature (see Figure 6–4) should keep inventory low and level.

**FIGURE 6–4**
**Modular Bill of Material**

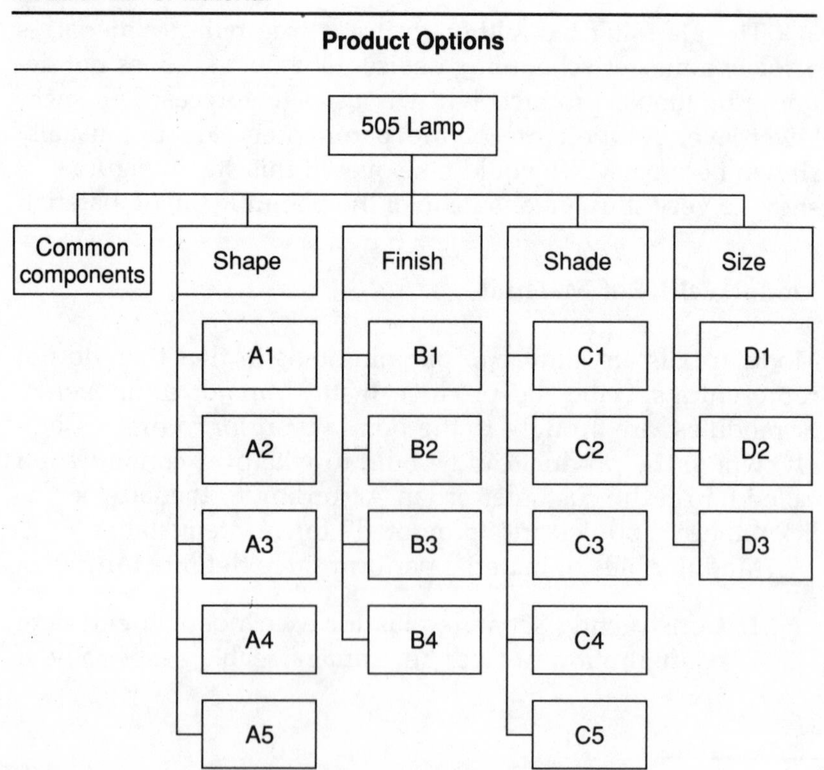


Possible combinations: 5 shapes × 4 finishes × 5 shades × 3 sizes = 300 variations
Minimum number of bills of material: 17 modules

### Pseudo-Bills of Material

This variation of the planning bill of material is also called a *super bill*. In the modularization process, parts can be logically grouped together because they use the same raw materials or components, perform the same or similar function, power steering versus manual steering or air conditioning versus none, etc. The objective here is to have the smallest number of items to forecast and master-schedule. These groups are assigned to an artificial parent (hence, *pseudo*) with a nonengineering part number. This is done for convenience only, for the bill does not represent a buildable unit. Rather, it is for planning and scheduling the least number of items.

The standard format for the bill of material is the single- or multilevel, indented type. The other four variations of the BOM have been included even though they are intended to handle unique, but recurring, situations.

One important feature that needs to be mentioned in regard to MRP is the item master file. This file is coded by part number and contains all pertinent permanent attributes of the individual part. The item master file can be as brief or as extensive as one needs. In addition, it may or may not include information regarding its current status. If your file does include status information, it is important that your computer program is capable of extracting just the current status when regenerating an inventory stock status report for planning and does not include a lot of extraneous data.

It is these item records linked together by their relationship to other (including parent) parts that develops in bill of material. The bill of material is the operating medium or the extension of the master scheduled item. It is vital it be correct. Attempting to operate via MRP with incorrect bills is an exercise in futility, and purification of the bills of material is imperative before embarking on an MRP system.

## INVENTORY STATUS

This is a given. Unless your records of inventory balances are correct and remain correct on an ongoing basis, your MRP system is going to fail. It is as simple as that. Moreover, there is only one way to accomplish this: by cycle counting. But all too often the responsibility for a system of cycle counting is given to an individual (more than likely an expeditor) in addition to his or her primary responsibilities. This is wrong! Cycle counting is a primary responsibility. The need for it is continuous and once this continuity is broken, you creep insidiously into the great morass of guessing whether the stock balances are correct. At this point you have lost control; once lost, only an unusual input of effort will revive it.

The treatment of perpetual inventory records is really a very elementary process. It entails the recording of transactions into and out of inventory. Inputs are any receipt of materials from vendors, shop orders, rework, and customer returns, as well as returns to stock. Inventory records must also include cycle count adjustments that show extra parts on hand. Outputs are any issues from stock. Mainly these are to support production, but other minor situations will cause pulls from stock from time to time. Here again, cycle counting may uncover decreases that must be adjusted. Allocations, or uncashed stock requisitions, represent an in-between situation, and their effect must be recognized. Discussion of allocations will be taken up shortly.

The only feasible way to handle transactions so that inventory records are not abused is through proper documentation. This cannot be stressed too strongly, as improper handling is what causes errors in the first place. Anytime there is an action upon an inventory part—be it used, moved, scrapped, or shipped—it is documented and fed into the system so that everyone is working with the same information. Traditionally this is the hardest nut to crack for most companies. Volumes have been written on how best to control it. One fact remains before all others: There is no purpose in even thinking about implementing MRP until this riddle has been solved.

## STATUS OF ON-ORDER PARTS

An important input to the MRP list is what is scheduled to come in from either vendors or manufacturing shop orders. The same rules stated for on-hand parts apply to on-order parts.

The data for lot sizing, lead times, and safety stock may be included in the item master file because they are not subject to frequent change. Nevertheless, a good bit of planning judgment comes into play here, for which we offer the following guidelines.

### Lot-Sizing Rules

The most common lot-sizing rules were discussed in an earlier section, but they can be summarized in three brief statements regarding common sense manufacturing:

1. By and large, lot-for-lot is the only logical way to go in an MRP environment.
2. Lot-for-lot can be enhanced by the use of the period order quantity (not the formula found in many texts, but the common sense rule).
3. Where specific situations suggest that definite benefits will accrue, either fixed-order quantity or fixed-period ordering may provide a viable alternative to either of the above rules.

### Lead Times

The determination of lead times is always a topic of concern, especially in view of the current interest in Just-In-Time manufacturing. Many texts explain in detail planned versus actual lead times and purchasing versus manufacturing lead times in an attempt to get a handle on this elusive subject, which impacts directly on both productivity and customer service.

Everyone knows from experience that lead times can vary from a day or two to infinity, depending upon (and, it seems, inversely proportional to) how desperate the need is for the particular item. What is at the heart of the matter is *realistic* lead times. Average or actual lead times tend to exacerbate the problem because by the time it is discovered that the material is late, the damage has already been done. That is why more time needs to be spent on planning to see whether one's expectations are realistic. Vendors can be made more responsive to actual needs when they know you are serious (and pay their bills on time). One constant source of amazement is how many companies religiously buy from the lowest-price supplier regardless of how unreliable their delivery history is.

Getting realistic lead times from vendors can be achieved by negotiation, but manufacturing lead times are something else entirely. When a master-scheduled item involves several operational or subassembly steps, the bill of material will be back-scheduled level by level to the beginning operation. These times added together give the total elapsed time. The path that requires the most time—the critical path—is the longest cumulative lead time. The length of the manufacturing cycle is determined by the longest cumulative lead time, which, in turn, will give the latest start date. Experience shows, however, that this is too ambitious. Some event or series of events is likely to occur to cause the completion date to be missed. This might be called the Murphy's Law of lead time. No one knows what will happen nor when, but something almost invariably does.

The function of cumulative lead time and its effects are depicted in Figure 6–5. Your attention is called to the administrative portion of lead time (dotted lines). Typically this runs from several minutes to perhaps as long as a week, rarely longer. All that is required here is a moment or two of the planner's attention to confirm the readiness of an order. This is the time an order sits on the planner's desk before it's released to the shop. The 2½ weeks allowed for is much too long, contingent events being what they are.

Based on experience, the planner will inflate the lead time to get a more realistic picture of probable elapsed time; excessive padding, however, should be avoided. Consider the other side of the coin. Suppose no hitch occurs and the finished parts breeze through in less than the scheduled time and some idle time is incurred. Is this bad? Let's face it: The goal is output—finished products out the door. This is where profitability begins (or ends).

What is gained by a process running at 120 percent or even 150 percent efficiency? All that results is that the load stays a

**FIGURE 6–5**
**Cumulative Lead Time**

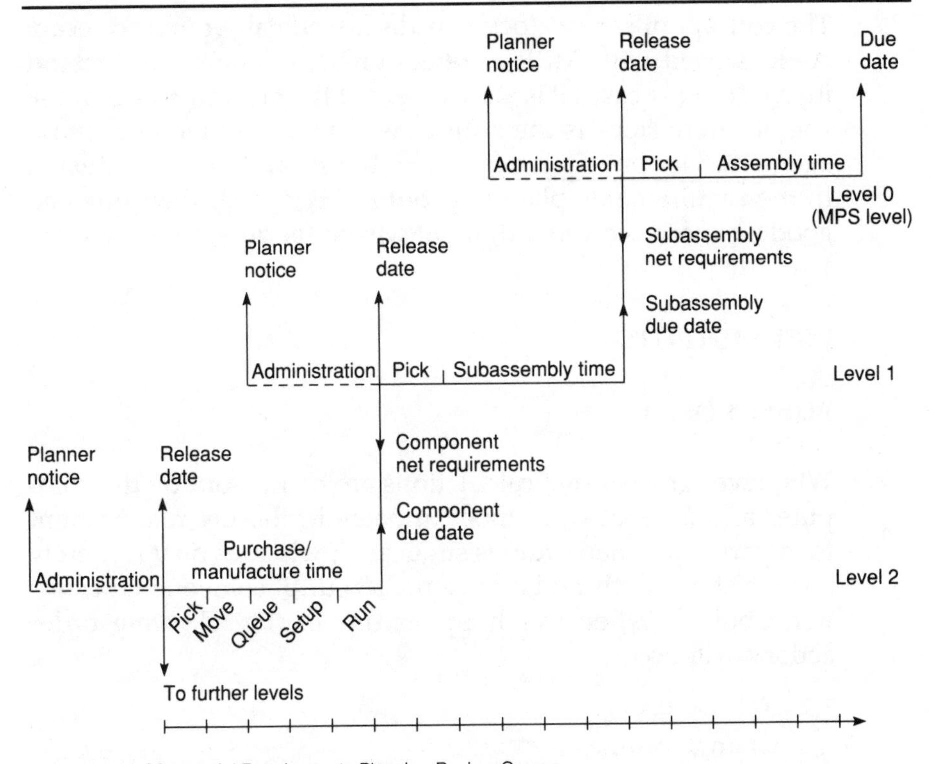


Source: APICS Material Requirements Planning Review Course

while longer in a queue at the succeeding work center. It does not add a single dollar to profitability.

That kind of efficiency is not the key—output is. The paramount objective is always the satisfaction of the needs of the customer. How it happens is of secondary concern, so if in the course of things some idle time occurs, it is no big deal. Companies who have moved toward Just-In-Time precepts have already discovered this. Just as the MPS must be ambitious, yet attainable, lead time must follow the same realistic logic.

### Safety Stock

The concept of safety stock was discussed earlier, but to recap: As it is used with MRP, it refers only to dependent-demand items. In some cases it is advisable, but the fact remains that the use of safety stock is the antithesis of modern inventory management. Not only is it expensive, but it can lead to confusion in the requirements-planning routine. For other than finished goods, experience and judgment will better determine its need.

## MRP OUTPUTS

### Planned Orders

Whenever gross-to-net calculations are carried out by the computer and a negative number appears in the net requirement for a part, a planned order is suggested, whereupon it is merely a suggestion, with no basis in reality, until someone takes action upon it. When this happens one of the following order actions will occur:

Release order.
Change order.
Replanned order.
Cancel order.
Exception report.
Firm planned order.

It would seem obvious that this is the most crucial part of the entire material requirements planning scheme. And it is this function that is most subject to the vagaries of Murphy's Law.

Some people describe an MRP system as sometimes being "nervous." Most of this nervousness, however, stems from their hesitancy to break away from older, more traditional concepts that never really worked but did provide ready excuses. Although planned orders and "on order" contribute most to nervousness, this doesn't tell the complete story. A strong suggestion of lot-for-lot requirements planning rule projected by the system no longer allows the luxury of in-excess-of-requirements inventory promoted by formula-driven concepts. This, along with the eradication of safety stocks, is certain to generate a good deal of planner nervousness. In addition, a great many of the occurrences that routinely plague a manufacturing shop—unmet shop and vendor lead times, shortages, and ever changing production schedules—all seem to dump on the planner's desk. It is not, nor will it ever be, an easy task.

## MRP Logic and Operation

The Master Production Schedule drives MRP. The bill of material is a reflection of the MPS broken down by individual part number. The bill of material, therefore, is the only source for determining gross requirements. The rule must be that never (and this means not ever, at any time) is anything involving production material removed from the premises without the direct intervention of the materials control department. Only they have the most complete and up-to-date information regarding the entire requirements planning picture. By removing one set of controls, the system begins to break down and it becomes easier, sometimes even necessary, to bypass other controls.

The basic function of MRP is to calculate the quantities needed to support production requirements (via the bill-of-material explosion) and time-phasing these requirements so that they are available when needed. The result of this computation is to cover the items on the schedule with a minimal amount of residual inventory to be carried forward into subsequent periods. This residuum is called *projected on hand* or *projected available inventory.* The objective is to keep this projected figure at zero, or as close to zero as is logical and reasonable. Any safety stock carried would have to be netted out because in an MRP system, safety stock is not available inventory.

The beginning number in this calculation is the gross requirements: the total quantity required for an individual part as spelled out by the bill(s) of material at any given time. The gross requirements less the inventory on hand and on order will yield the net requirements. It is assumed, of course, that the on-hand and on-order figures have had any prior allocation netted out so that the net requirements reflect the true status for planning purposes.

If the calculated net requirements are zero or above, no action need be taken. Any residuum can be carried forward to the next planning period and the process repeated. If the requirements fall into the minus category, an order must be placed to bring the item back into balance with the requirements. This is the output function of MRP and is called *planned orders* and *planned order releases*.

Since material requirements planning is a computer-based system, there are several ways information can be formatted using the time-phased concept. One is called *date/quantity*, where the data are displayed vertically with each day and its associated transaction on a separate line. This is processed automatically, and a new projected-on-hand is calculated daily. Another format is the *time bucket* approach, where the data are set out horizontally in a progressive series of time periods, normally a week, with a separate transaction number for each line. Most computer programs are formatted with the date/quantity approach because it is more accurate and more easily updated and maintained to keep priorities and due dates valid.

Time buckets displayed on a CRT provide a useful planning tool because the relationships of various input and output transactions are immediately shown. Some users suggest combining the two formats. The date/quantity would be the active file, while particular activities and functions would be planned using time buckets. With the bucket period usually being a week, timing of specific events may become important. This would be handled by the date/quantity printout, where all the categories are updated daily. Typical formats of the two approaches are shown in Figures 6–6 and 6–7.

The logic of the two types of formatting are readily apparent, but the operation of the system requires some explanation:

1. Gross requirements appear only in the period in which they are due. They are not accumulated.
2. The visibility provided by the time buckets over the planning horizon allows requirements to be planned in advance of their need with lead time offset (setback)

**FIGURE 6–6**
**Time Bucket Format**

| Lead time: 2 weeks | | Week 14 | 15 | 16 | 17 | 18 |
|---|---|---|---|---|---|---|
| Gross Requirements | | 0 | 50 | 100 | 0 | 20 |
| Scheduled Receipts | | | | 150 | | |
| Projected on Hand | 50 | 50 | 0 | 50 | 50 | 30 |
| Planned Order Release | | (150) | | | | |

**FIGURE 6–7**
**Date/Quantity Format**

| *Week* | *Work Day* | *Transaction* | *Quantity* | *Projected on Hand* |
|---|---|---|---|---|
| 14 | 93 | | | 50 |
| 14 | 93 | Planned Order | 150 | 200 |
| 14 | 94 | Planned Order Requirements | 150 | 200 |
| 15 | 100 | Gross Requirements | 50 | 150 |
| 16 | 107 | Gross Requirements | 50 | 100 |
| 16 | 108 | Scheduled Receipts | 150 | 100 |
| 16 | 110 | Gross Requirements | 50 | 50 |
| 18 | 122 | Gross Requirements | 20 | 30 |

from the date of the scheduled receipt as the arrow beneath the time bucket display indicates.

3. The ability to visualize lead time offset sets the time bucket approach apart from date/quantity.
4. In the date/quantity example, the projected on hand is increased as the planned order is generated instead of when the order is released or received. This is not a hard-and-fast rule; your program will dictate the way this is handled.
5. The planned order for 150 is in parenthesis (time bucket display) because the order will show as either a

**FIGURE 6–8**
**Unplanned Requirements: Time Bucket**

| | | Week | | | | |
|---|---|---|---|---|---|---|
| | | 14 | 15 | 16 | 17 | 18 |
| Gross Requirements | | 0 | 50 | 100 | 0 | 20 |
| Scheduled Receipts | | | | | | |
| Projected on Hand | 50 | 50 | 0 | -100 | -100 | -120 |
| Planned Order Release | | | | | | |

planned order or a scheduled receipt, but never both. Once the planned order is released, it becomes a scheduled receipt. Both are shown in the example for demonstration purposes only. The unplanned time bucket display would appear as in Figure 6–8.

This example indicates a net requirement of at least 120

6. The time bucket display must have *at least* four transaction categories. Obviously, it can have more but will always include these four.
7. A figure always appears for the projected on hand line under either format.
8. A good rechecking technique can be utilized to ensure you are in balance: The total of gross requirements less the total of scheduled receipts less beginning inventory plus ending inventory will equal the total of planned orders.
9. If at any time during either the planning cycle or the action cycle the projected on hand for a part drops to zero (or a very low number), it becomes a reasonable candidate for cycle counting. In this way more parts can be counted or fewer hours devoted to it.
10. In respect to the date/quantity format, it's worth noting that in order to preserve chronology the report is printed in the time sequences in which transactions take place. It would appear that the planned order was

**FIGURE 6–9**
**Unplanned Requirements: Date/Quantity**

| *Week* | *Work Day* | *Transaction* | *Quantity* | *Projected on Hand* |
|---|---|---|---|---|
| 14 | 93 | | | 50 |
| 15 | 100 | Gross Requirements | 50 | 0 |
| 16 | 107 | Gross Requirements | 50 | -50 |
| 16 | 110 | Gross Requirements | 50 | -100 |
| 18 | 122 | Gross Requirements | 20 | -120 |

generated independently of the requirements it was meant to cover, but this obviously is not the case. The planned order is forced into its slot because of the lead time offset. The unplanned date/quantity printout would appear as in Figure 6–9.

This process is repeated for every inventory item involved in the manufacturing cycle over the planning horizon. Special attention must be paid to those parts whose lead time extends beyond the horizon. The next operational step carries this process on completion. Planned orders released at one level become gross requirements at the next level down, unless they come from a vendor. This relationship of parent to lower-level item extends throughout the system. As the bill of material is exploded level by level, it gives rise to a great number of subparts, each with its own set of attributes regarding lead time, procurement, and processing. (This is the central nervous system of Material Requirements Planning.)

The entire MRP scheme is detailed in Figure 6–10.

## SPECIAL SYSTEM FEATURES

Because MRP was developed as a computer-controlled system, several things evolved that have become primary tools to aid the planner in system management.

**FIGURE 6–10 Material Requirements Planning Scheme***

| Lot Size | Lead Time | On Hand | Safety Stock | Allocated | Low Level Code | Item Number | | | Period 1 | 2 | 3 | 4 | 5 | 6 | 7 | 8 |
|---|---|---|---|---|---|---|---|---|---|---|---|---|---|---|---|---|
| 25 | 2 | 20 | — | — | 1 | | Gross requirements | | 10 | | 5 | 15 | | 5 | 20 | 5 |
| | | | | | | | Scheduled receipts | | | | | | | | | |
| | | | | | | | Projected on hand | 20 | 10 | 10 | 5 | –10 | | –15 | –35 | –40 |
| | | Parent Assembly — A | | | | | Projected available | | 10 | 10 | 5 | 15 | 15 | 10 | 15 | 10 |
| | | | | | | | Net requirements | | | | | 10 | | | 10 | |
| | | | | | | | Planned order receipts | | | | | 25 | | | 25 | |
| | | | | | | | Planned order releases | | | 25 | | | 25 | | | |
| 50 | 4 | 60 | 5 | — | 2 | 1 | Gross requirements | | | 25 | | | 25 | | | |
| | | | | | | | Scheduled receipts | | | | | | | | | |
| | | | | | | | Projected on hand | 60 | 60 | 35 | 35 | 35 | 10 | 10 | 10 | 10 |
| | | Component 1 | | | | | Projected available | | 60 | 35 | 35 | 35 | 10 | 10 | 10 | 10 |
| | | | | | | | Net requirements | | | | | | | | | |
| | | | | | | | Planned order receipts | | | | | | | | | |
| | | | | | | | Planned order releases | | | | | | | | | |
| 30 | 1 | 10 | — | — | 2 | K | Gross requirements | | | 25 | | | 25 | | | |
| | | | | | | | Scheduled receipts | | | 30 | | | | | | |
| | | | | | | | Projected on hand | 10 | 10 | 15 | 15 | 15 | –10 | | | |
| | | Component K, a Subassembly | | | | | Projected available | | 10 | 15 | 15 | 15 | 20 | 20 | 20 | 20 |
| | | | | | | | Net requirements | | | | | | 15 | | | |
| | | | | | | | Planned order receipts | | | | | | 30 | | | |
| | | | | | | | Planned order releases | | | | | 30 | | | | |
| | | | | | | | Gross requirements | | | | | 30 | | | | |

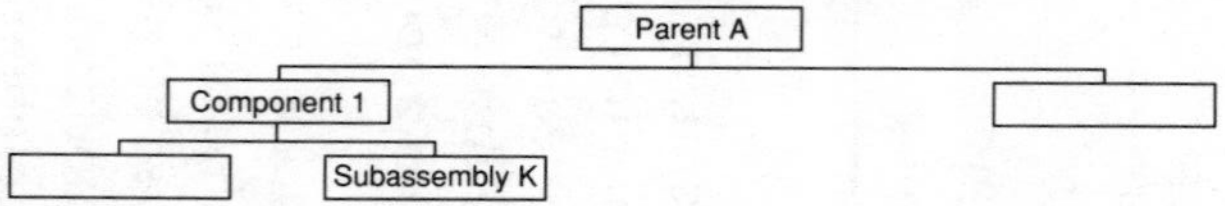

*The planned-order releases of 25 parent assemblies become gross requirements at the next level down, component 1. Because there are enough component 1's to satisfy this requirement, it passes through to its subparts, one of which is subassembly K. This subassembly has some stock and enough coming in to fulfill the first requirement but not the second, so an order is planned for release in week 4. Assuming subassembly K has its own set of dependent parts, the same iterations would continue down to their lowest levels.

Source: APICS Material Requirements Planning Review Course

## Firm Planned Order

Whenever a schedule change forces a system into a replanning mode, the computer will automatically reallocate existing and planned orders to different (mostly later) time periods. The purpose of the firm planned order (FPO) is to prevent this from happening. The program logic must be capable of coding an order to preclude it from being automatically replanned. There are various reasons for the firm planned order, but primarily it is used to dampen change in the near term, when shifting the sequence of orders on the production floor would be extremely disruptive and impact negatively on inventory levels. Under conditions of use, the FPO is treated just like an open order or scheduled receipt, with an exception report sent to the planner. Unless the planner intervenes, the FPO, when it matures, will enter the production stream as originally scheduled. This feature is most helpful when a change is suggested inside of cumulative lead time, when an adverse effect on customer service is probable.

## Pegging

The only difference between pegging and the where-used file is that the latter is passive while pegging is active. That is why pegging is termed a *live-where-used* or *selective-where-used file*. What pegging does is call out or tie a component to its parent at the next level up. This is single-level pegging. If the planner wishes to inquire about parent assemblies further up the product chain, the system must have "full-peg" capabilities. This capability aids the planner by providing answers to problems as they arise. An example will show how this works.

Assume a product goes through four levels of operation up to the completion (end item) level. The order is for 100 finished parts that are built up in part from a like quantity of a certain raw material. As the parts are staged for production, it is discovered that of the 100 units of raw material received, 50 were rejected by Receiving or Incoming Inspection and are unusable. Faced with this quandary, the planner wishes to find out if enough residual parts remain at the various levels to make up all, or at least a substantial part of, the requirements and so uses the pegging procedure to accomplish this. The raw material is pegged to its immediate parent part on level four, and it is found that nine parts remain from a previous run. Used as a dependent part, it is pegged to its parent on level three, where 22 more are found. Repeating the process for level two reveals 14 more; for level one an additional two. All these parts are added together to produce this schedule:

| | |
|---|---|
| Level 4 | 9 |
| Level 3 | 22 |
| Level 2 | 14 |
| Level 1 | 2 |
| Total | 47 |

which added to the 50 received makes an available total of 97. This is close enough, so production proceeds as scheduled. Figures 6–11 through 6–18 illustrate the sequences as the items are pegged level by level.

The customer places an order for 100 units of Product 0, which has been master-scheduled. Total cumulative lead time is seven weeks, and Product 0 is to ship in week 21. The B/M is exploded to its lowest level, which reveals need for sufficient raw material of Part 505 to make 100 units. The planned order is drawn and scheduled for release in week 14 as shown in Figure 6–11.

**FIGURE 6–11**
**Pegging Sequence (1)**

| Part number 505<br>Lead time 2 weeks | | Week<br>14 | 15 | 16 | 17 | 18 | 19 | 20 | 21 |
|---|---|---|---|---|---|---|---|---|---|
| Gross Requirements | | | | 100 | | | | | |
| Scheduled Receipts | | | | | | | | | |
| Projected on Hand | 0 | | | | | | | | |
| Planned Order Release | | 100 | | | | | | | |

**FIGURE 6–12**
**Pegging Sequence (2)**

Part number 505
Lead time 2 weeks

| | Week | | | | | | | |
|---|---|---|---|---|---|---|---|---|
| | 14 | 15 | 16 | 17 | 18 | 19 | 20 | 21 |
| Gross Requirements | | | 100 | | | | | |
| Scheduled Receipts | | | 100 | | | | | |
| Projected on Hand | | | | | | | | |
| Planned Order Release | | | | | | | | |

The order now becomes a scheduled receipt due in week 16. Material has been received, but only enough for 50 units. An exception report (XR) shows the balance rejected by incoming inspection. Because the due date is now within the cumulative lead time, the planner cannot wait for another receipt. A limited number of options are available:

Other available material must be found.

A suitable substitute, if any is available, can be used.

Half the order can be run.

One of the above must be done

Another alternative exists, but it should be avoided if at all possible: getting an emergency replenishment of the raw material and pushing it through the shop manually, making it a priority item at each work station. This causes extreme disruption; however, it's better to make the best of a bad situation and educate the supplier on the possible consequences of a repeat performance.

The Planner searches the file to find the parent part that uses RM 505. He does this be determining which parent using 505 has an active gross requirement against it. (Those cases that show more than one active requirement outstanding will be taken up later.) Part 404 is found to be an active parent of Part 505. Nine pieces are shown to be in inventory.

The next level up is scanned, repeating the same process, and Part 303 is uncovered. Note the reversal of the system logic where planned orders at one level become gross requirements

**FIGURE 6–13**
**Pegging Sequence (3)**

Part number 505
Lead time 2 weeks

| | | Week 14 | 15 | 16 | 17 | 18 | 19 | 20 | 21 |
|---|---|---|---|---|---|---|---|---|---|
| Gross Requirements | | | | 100 | | | | | |
| Scheduled Receipts | | | | | | | | | |
| Projected on Hand | 0 | | | 50 | XR | | | | |
| Planned Order Release | | | | 50 | | | | | |

**FIGURE 6–14**
**Pegging Sequence (4)**

Part number 404
Lead time 1 week

| | | Week 14 | 15 | 16 | 17 | 18 | 19 | 20 | 21 |
|---|---|---|---|---|---|---|---|---|---|
| Gross Requirements | | | | | 100 | | | | |
| Scheduled Receipts | | | | 50 | | | | | |
| Projected on Hand | 9 | 9 | 9 | 59 | 0 | | | | |
| Planned Order Release | | | | | 59 | | | | |

at the next level down. Here the planned orders reflect fulfillment of the gross requirements at the next level higher, only moved forward as the result of the one-week lead time. Here the inventory shows 22 pieces available plus the 9 found before allow 81 parts to be assembled at the next processing level.

The following illustrations show successive iterations of the pegging process up to the shipping point.

Caution must be exercised in two areas. The first is obvious: The parts must be counted to ensure that they actually exist. Inventory records (believe it or not) have been known to be wrong. The second is not at all obvious. It is entirely possible, especially if these are common parts, that they have been planned to cover another, different requirement. This can be a difficult problem, and how it is solved is covered in the next section.

**FIGURE 6–15**
**Pegging Sequence (5)**

Part number 303
Lead time 1 week

| Week | | 14 | 15 | 16 | 17 | 18 | 19 | 20 | 21 |
|---|---|---|---|---|---|---|---|---|---|
| Gross Requirements | | | | | | 100 | | | |
| Scheduled Receipts | | | | | 59 | | | | |
| Projected on Hand | 22 | 22 | 22 | 81 | 0 | | | | |
| Planned Order Release | | | | | | 81 → | | | |

**FIGURE 6–16**
**Pegging Sequence (6)**

Part number 202
Lead time 1 week

| Week | | 14 | 15 | 16 | 17 | 18 | 19 | 20 | 21 |
|---|---|---|---|---|---|---|---|---|---|
| Gross Requirements | | | | | | | 100 | | |
| Scheduled Receipts | | | | | | 81 | | | |
| Projected on Hand | 14 | 14 | 14 | 14 | 14 | 95 | 0 | | |
| Planned Order Release | | | | | | | 95 → | | |

# MANAGING SYSTEM RESPONSIVENESS

## Frequency of Replanning

Because the routine of manufacturing is in a constant state of flux, with inputs and outputs occurring continually, the question arises as to how often the records should be updated—or purified, as it were. Inherent in the MRP system is the perpetual need to keep priorities and due dates valid. The validity problem is endemic to the timeliness of the record updating relative to its feedback, or response. What, when, and how often to communicate changes are the decisions a company has to make. As with every decision, they are tied to the time and

**FIGURE 6–17**
**Pegging Sequence (7)**

| Part number 101 Lead time 1 week | | Week 14 | 15 | 16 | 17 | 18 | 19 | 20 | 21 |
|---|---|---|---|---|---|---|---|---|---|
| Gross Requirements | | | | | | | | 100 | |
| Scheduled Receipts | | | | | | | 95 | | |
| Projected on Hand | 2 | 2 | 2 | 2 | 2 | 2 | 97 | 0 | |
| Planned Order Release | | | | | | | | 97 | |

**FIGURE 6–18**
**Pegging Sequence (8)**

| End item 0 Lead time 1 week | | Week 14 | 15 | 16 | 17 | 18 | 19 | 20 | 21 |
|---|---|---|---|---|---|---|---|---|---|
| Gross Requirements | | | | | | | | | 100 |
| Scheduled Receipts | | | | | | | | | |
| Projected on Hand | 0 | | | | | | | | 97 |
| Planned Order Release | | | | | | | | | -3 |

the cost of processing MRP. The choice is whether to regenerate all the records on a periodic basis or to change only the affected categories each time a transaction occurs.

## Regeneration

The complete current master production schedule is run and the respective bills of material are exploded from top to bottom, level by level. The status of each active inventory item is brought up-to-date. Any changes that affect either requirements or inventory after a run must wait for the next regeneration.

## Net Change

Only those transactions that cause a change in the status of a master-scheduled item are exploded. The existing schedule is viewed as one continuous plan with successive changes.

Of the two types of file updating, regeneration is obviously the more favorable, although it may not always be the more feasible. In a time-driven system like regeneration, any changes in the priority status are not reflected until the next run comes due. That is why regeneration must take place at least once a week. Net change, on the other hand, is transaction-driven. Anytime a change occurs affecting requirements or inventory, it triggers a partial explosion.

There are two special considerations regarding file updating that must be mentioned: item record balance and intralevel equilibrium. The former balances net requirements against planned orders. This, plus the projected on-hand, must equal gross requirements. Intralevel equilibrium takes into consideration that a transaction at any level may affect the status of lower-level dependent requirements and that these gross dependent requirements must always equal the planned orders generated by the parent items both in quantity and timing.

What this means is that the unique processing logic of the MRP system will automatically maintain item record balance for planned transactions. A planned input into the system causing an imbalance will immediately trigger an offsetting transaction somewhere else, bringing the item records back into balance (e.g., planned orders becoming scheduled receipts or scheduled receipts becoming projected on-hand). During this process, intralevel equilibrium would remain unaffected. However, it is unplanned transactions that will disturb this balance and upset the status of both. Unplanned withdrawals—issues from stock, scrap, or changes, plus or minus, resulting from cycle counting, as well as schedule changes—will force the scheduler to analyze the transaction and its effect on the system. This may bring about a new explosion and a rescheduling of the affected areas, where necessary, to rebalance the system.

For example, look at Part 101 in Figure 6–18. Tracking down the two parts showed that they had been scrapped sometime earlier. The records must be changed to reflect this, but there is no automatic rebalancing transaction code that will do it. Because this is an unplanned transaction, intralevel equilibrium will never be realized unless the planner intervenes directly and forces the change manually. Once this is completed, the necessary categories will be updated to reflect actual status. Had the scrappage occurred further downstream—say, in Part 303—the change would have been forced here and a new (partial) explosion run to show how all the levels above 303 were affected by the change.

On the other hand, when the 50 pieces of RM 505 were rejected, this transaction, although obviously unplanned, brought itself into item record balance automatically. Because the scheduled receipt was not received complete, a new planned order displayed the difference, suggesting some action by the planner. Referring to Figure 6–14, this is precisely what happened.

As a matter of fact, any number other than 100 will show up in the planned-order release column because the item records must remain in balance.

Status before and after scrap is shown in Figures 6–19.

The net change, being transaction-driven, will necessarily force the scheduler to review each transaction, whatever its source. This will provide the impetus for maintaining item record balance and intralevel equilibrium. Regeneration, on the other hand, with periodic batch processing, is oblivious to these relationships.

Whether net-change transactions are processed on line or held and batch-processed daily, it will consume more total computer time than does regeneration. Its impact, however, is not as great because it is spread evenly over a longer period.

The choice between the two systems, if one can be made, is not merely the selection of regeneration or net change, one over the other. No one is going to choose net change as a feedback tool unless the particular operation fairly demands it. Many firms stay with regeneration when they should be oper-

**FIGURE 6–19**

Status before Scrap

| Part number 101<br>Lead time 1 week | | Week<br>14 | 15 | 16 | 17 | 18 | 19 | 20 | 21 |
|---|---|---|---|---|---|---|---|---|---|
| Gross Requirements | | | | | | | | 100 | |
| Scheduled Receipts | | | | | | | 95 | | |
| Projected on Hand | 2 | 2 | 2 | 2 | 2 | 2 | 97 | 0 | |
| Planned Order Release | | | | | | | | 97 | |

Status after Scrap

| Part number 101<br>Lead time 1 week | | Week<br>14 | 15 | 16 | 17 | 18 | 19 | 20 | 21 |
|---|---|---|---|---|---|---|---|---|---|
| Gross Requirements | | | | | | | | 100 | |
| Scheduled Receipts | | | | | | | 95 | | |
| Projected on Hand | .0 | | | | | | 95 | | |
| Planned Order Release | | | | | | | | 95 | |

ating under net change and solve the problem by running a complete regeneration several times a week, in which case processing time (not to mention cost of computer paper) becomes a major consideration. Moreover, practically all software packages available for micro- and minicomputers offer regeneration as the only option.

Net change is by no means a panacea. As stated earlier, it can only be a more feasible alternative. The longer the period of time the net-change file itself must be regenerated, the more it is subjected to deterioration. Errors will creep in and dampen its effectiveness and it will become a prime cause of nervousness.

Finally, with respect to pegging, it is easy to see that either of these two approaches will reveal any prior requirements, if they exist.

Now that the master production schedule and material requirements planning have been examined in detail, we present MRP II in Figure 6–20.

**FIGURE 6–20**
**Closed-Loop MRP (MRP II)**

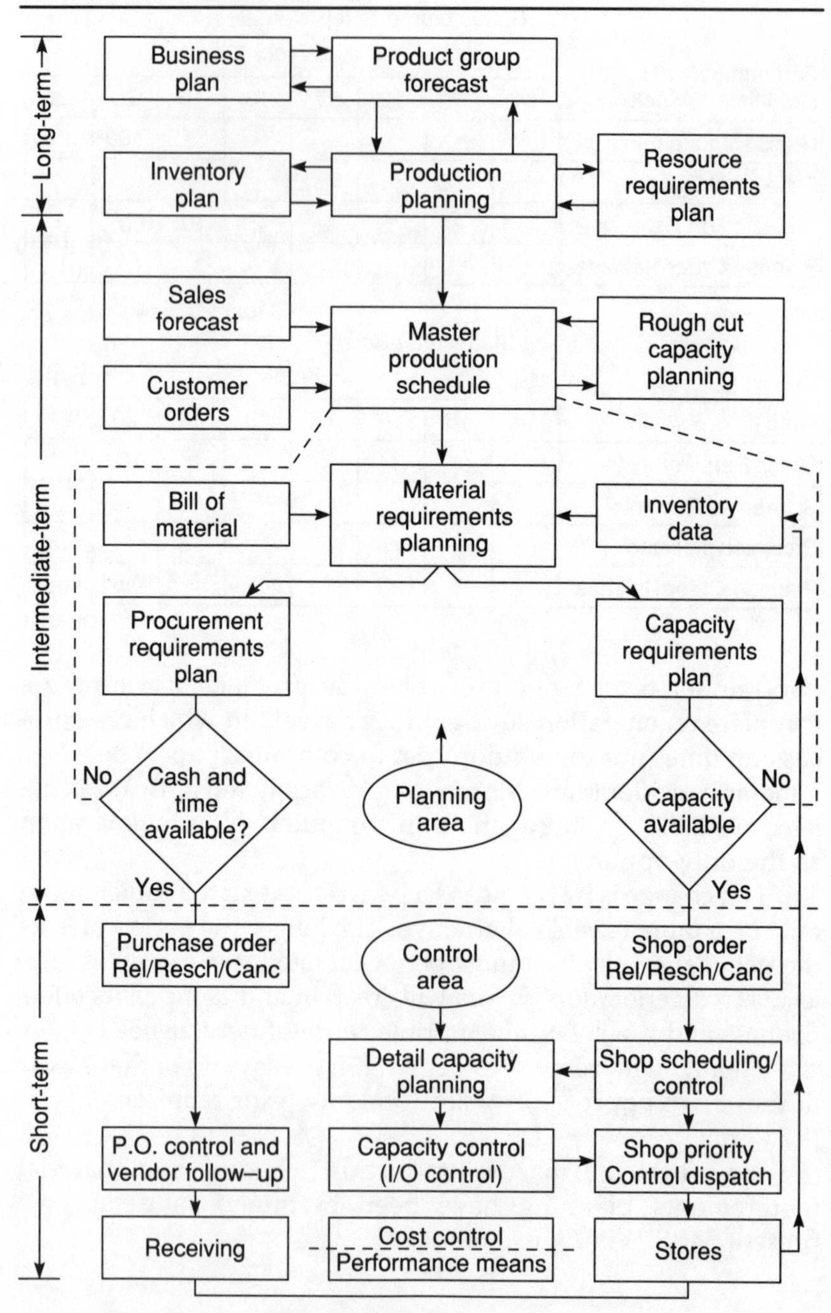


Source: Material Requirements Planning Review Course

## SUMMARY

Material requirements planning (MRP) is a living system. Combined with the master production schedule (MPS) it forms the "drive train" that powers manufacturing. The key to this is the incorporation of time-phasing, which has obviated the need for older and certainly less effective systems and techniques.

Sir Winston Churchill once said, that "democracy is the worst form of government except all those other forms that have been tried from time to time." The same can be said of MRP II as it relates to manufacturing management.

If these statements are held to be essentially accurate, the question now becomes: How has MRP performed over the breadth of American industry where it has been implemented? Unfortunately, and perhaps surprisingly, the record is not a good one. In a survey conducted several years ago it was found that slightly less than 10 percent of the companies surveyed reported that the system delivered as advertised. For those who reported positive results, however, the benefits were substantial. This does not mean that MRP is great in theory but doesn't work in practice. What *seems* to be at fault is system shortcomings, but it is a little more complex than that. There are no shortcomings with MRP; the shortcomings are people-related and manifest themselves in one or both of two ways.

First, there is no clear constancy of purpose about what is actually to be accomplished, of precisely what problems need to be resolved. Second, there seems to be a compulsion by otherwise intelligent and successful business people to fit neat little formulas into a system that was expressly designed to exclude them.

Material requirements planning is and does exactly what the name suggests: It provides management with an excellent planning tool. It is the responsibility of manufacturing to execute that plan.

# CHAPTER 7

# CAPACITY PLANNING AND CONTROL

## CAPACITY PLANNING

Capacity considerations are the most underutilized and least understood of all the various concepts and techniques that go toward making the material requirements planning system viable. This is probably due primarily to the fact that the methods used in setting production capacities are mercurial ones. Here, the subject of capacity deals with a great many unknowns. By and large, one almost has to wait for something to happen before he or she can react.

### Understanding Capacity

Knowing what your capacity is and how to deal with the constraints it imposes is not merely desirable, it is mandatory. The implications of capacity are just as inextricably interwoven with the primary goal of customer service as are any of the systems covered up to this point. It is the culmination of the production planning and control procedures. Thus far it has been determined what the customer wants and when, and the material has been acquired. Now we must find out if the plant has the wherewithal to get it done.

One condition must be understood before proceeding: MPS and available capacity must be in sync; that is, there must be sufficient capacity available to accomplish what has been directed by the MRP. If there is not, there are only two options: Increase capacity or reduce the schedule.

Historically, capacity considerations have not been dealt with extensively. More recently, however—because they are such an essential part of the MRP system—they have been receiving a lot more recognition. The concept of capacity has been divided into two broadly defined categories, capacity planning and capacity control, and that is how we will treat the subject here.

### Infinite versus Finite Capacity

A final point bears mentioning. The age-old argument about loading to infinite capacity versus finite capacity has finally been put to rest after a long battle. When each was dissected, it was discovered that, instead of being on opposite sides of the capacity coin, they were actually complimentary to one another. It was their names that misled people. Infinite capacity, because it encompassed a broader horizon, took on the name capacity requirements planning, while finite capacity was given the narrower definition of work center loading or operations sequencing.

### Capacity Requirements

The body of knowledge concerning capacity and its implications was further refined and enhanced when it was coupled with the MRP system. In the days preceding MRP—and to a great extent, even today—capacity was planned and controlled by the use of the machine load report. The work in the shop was scheduled onto the various work centers, and new orders were routinely fed in. How busy you were and how healthy you felt were a direct reflection of the order backlog. This was an extremely short-sighted approach and almost totally ineffective because it resulted in late orders, furious expeditors, and ever increasing lead times. The machine load report had no visibility; it provided no insight into those orders that needed to be planned and only accounted for those on the immediate horizon. It was, in fact, no more than a tallying sheet.

In the very near term, however, the machine load report did a great job where checking for underloads and overloads was necessary or for getting a handle on past-due orders and their aging. Beyond this, the machine load report did nothing about relating both planned and released orders to available capacity.

Therefore, a new scheme was required for in capacity management, a new set of procedures were needed for estimating, scheduling, and regulating capacity. Each component was reexamined and categorized, and a new system for planning and controlling capacity was fashioned that provided continuity with material requirements planning.

There are two things that make up capacity: time and labor input. The product of these two factors interacting in a particular production facility on a given part provides a basis for measurement and therefore capacity planning.

- CAPACITY PLANNING
  - Resource Planning
    - Product load profile
  - Rough Cut Capacity Planning
  - Capacity Requirements Planning
  - Finite Capacity Considerations
    - Machine loading
    - Operations sequencing
- CAPACITY CONTROL
  - Input/Output Control
    - Critical ratio

## RESOURCE PLANNING

Long-range capacity planning is necessarily an integral part of a company's overall business plan as it plots its strategy several years into the future. Companies continuously strive to meet their customers' needs, but they must balance this demand with their overall resources. Unless solid planning is undertaken at this stage, increased production costs and disruption of service could well result.

Whether you are operating with a conventional or Just-In-Time manufacturing strategy, the same capacity constraints exist. They may be allocated differently, but available capacity is a constant and the volume of production remains the same.

What is necessary is to construct load profiles for the product that will enter the manufacturing stream.

**Product Load Profiles**

In order to measure the impact on capacity of any given product mix in the MPS, a standard for measurement is needed. This is supplied by product load profiles.

Computing load profiles for individual products is based on the simple proposition that each quantity of product in the MPS generates a measurable load and that the procedures used to arrive at a machine load report can be used to compute a product load profile. Machine load reports should be part of the production records, but if they are not complete, estimates from recent runs must be made. Based on these numbers, a given load profile will consist of the standard hours required for the setup and fabrication for one unit of product on any stated machine or in a given process.

For example, assume the order calls for 1,000 pieces of a part that is scheduled to run at a particular work center. Setup time is estimated at 3.5 hours, and recent run reports show an average output of 133 pieces per hour.

1,000 ÷ 133 = 7.5 hrs. fabrication time
3.5 hrs. setup
11.0 hrs. processing time
11 hours = 660 minutes
660 minutes ÷ 1,000 = .66 minute
1 piece takes .66 minute, or 40 seconds
Cycle time = 40 seconds per piece

This calculation of the product load profile has two advantages:

1. Once constructed, it is not subject to very much change over time. The cycle time and setup time should become part of the item master file. If the run size should change, however, a new load profile will have to be recalculated.
2. It is the only definable way of measuring capacity, although a certain amount of caution should be exercised because the times are bench marks only.

This computation of the product load profile can be repeated for each component item appearing in the bill of material called out by the MPS.

Using a simulated lot size of one, all order quantities can be extended and summarized to show the total load on the factory for any specified time period. (The end result will be to determine the overall effect of the total Master Production Schedule.) Any overload or underload that develops will be clearly evident. By simulating alternative schedules, one can select the best proposed solution.

As these load profiles become part of the permanent planning records, data will be made available to complete the next step in the capacity-planning hierarchy.

## ROUGH-CUT CAPACITY PLANNING

In order to avoid disruptive and costly changes in the master schedule, one must investigate bottleneck situations that may develop at critical work stations. First it is necessary to identify the load upon these work centers as the schedule progresses through time.

This identification process is known as rough-cut capacity planning. It uses input from the master production schedule and covers the same time horizon as the schedule does. Its purpose is specific: Before delivering the proposed master schedule to MRP for planning quantities and dates of requirements, it must first be determined that sufficient capacity is available, that the MPS is makeable.

Product load profiles are used for this purpose. A load profile, extended by the scheduled quantity, will indicate the total requirements upon a given process and whether there is sufficient capacity for the time period involved. If sufficient capacity does not exist, to repeat, some orders will have to be scheduled at a later date or the scheduled quantities will have to be reduced.

Obviously it would be unwise to perform a detailed capacity check at this point, because this is always done at the next level down. Only those processes that show a chronic overload and impact negatively on customer service need to be examined.

## CAPACITY REQUIREMENTS PLANNING

Once the MPS is compiled in final document form (published), it becomes the instrument for planning by MRP. Intrinsic to the process of planned orders, the output portion of MRP, is that open capacity exists to produce orders within the framework of cumulative lead time—they can be procured when they are needed. In other words, when orders are being generated, they are specific only in relation to quantity and time, without regard for sufficient available capacity. Only when an order is assigned to a specific process does the question of capacity arise. Does enough open capacity still exist on the machine or at the work center at the time scheduled to run the amount required?

Capacity can be measured in either of two ways. Using the quick-and-dirty method, you can take the units run or hours earned during a week for a particular operation over a series of weeks and obtain an average output, to which you must add an intuitive factor and use the result as a guide.

A better way is to calculate the capacity of a particular process, which involves several steps. First, you have to know what the normal capacity of the work center is. This is the number of shifts you run per day times either the number of operators or the number of machines operated times the standard number of shift hours times the length of the workweek. For example,

| | |
|---|---|
| 2 | shifts |
| X2 | operators (or machines) |
| X8 | shift hours |
| X5 | week days |
| 160 | hours of normal capacity |

It's obvious that getting 160 hours out of the department in any week is an impossible task, so this figure has to be tempered with utilization and efficiency factors.

Utilization is the percentage of elapsed time a work center is active. This is obtained by dividing scheduled hours into the number of hours actually used. *Scheduled hours* is the key term here. Not included is time spent on lunch, breaks, time when an operator is not available, and the like. Scheduled hours are those when the equipment should have been running but wasn't due to the lack of material or the need for maintenance, repair, etc. Obviously utilization can never exceed 100 percent, which would be more hours than there are available.

However, efficiency, which is simply the ratio of actual output to planned output, can be less than or greater than 100 percent.

Several facts must be realized about both utilization and efficiency: (1) The numbers used for comparing them must be stored in the computer; (2) the numbers must be expressed as percentages; and (3) the numbers that contribute the input are not important by themselves; it is the percentages calculated from them that must reflect reality.

Once the percentages are calculated, the utilization/efficiency relationship will be their product. For example, if you determine a 90 percent utilization, the machine was actually churning out parts 90 percent of the scheduled time. If you have 90 percent efficiency, 90 percent of the planned amount went into inventory at the end of the scheduled period. The product of these percentages, 81 percent, multiplied by the rated (normal) capacity will be the expected standard hours that can be planned against a work center.

If utilization is 80 percent (instead of 90 percent) and efficiency is 110 percent (instead of 90 percent), the equation for the utilization/efficiency factor will be:

$$80\% \text{ utilization} \times 110\% \text{ efficiency} = 88\%$$

Applying the formula to both examples, we obtain

| 2 shifts | 2 |
|---|---|
| X2 operators (or machines) | X2 |
| X8 shift hours | X8 |
| X5 week days | X5 |
| X81% utilization/efficiency factor | X88% utilization/efficiency factor |
| 129.6 standard hours | 140.8 standard hours |

An additional, unintended benefit of determining the utilization and efficiency factors is that the analysis involved in the calculations assists management in evaluating short-term capacity problems and provides a basis for future improvement.

One final item can be added to this capacity measurement formula for the weekly output a work center should produce: overtime. Overtime is also expressed as a percentage. If it is scheduled for a particular operation at one extra hour per operator, per shift, per day, it would increase capacity by 12.5 percent:

$$(5 \text{ hours/week} \times 2 \text{ shifts} \times 2 \text{ operators}) \div 160$$
$$= 20 \div 160 = 12.5\%$$

If this overtime is added to the two previous examples, we obtain

$$129.6 + (129.6 \times .125) = 129.6 + 16.2 = 145.8 \text{ hours}$$

and

$$140.8 + (140.8 \times .125) = 140.8 + 17.6 = 158.4 \text{ hours}$$

Caution must be exercised here because, as a rule, overtime hours are not as productive as regular hours, and adding still more overtime merely compounds the problem.

Constructing capacity profiles for the entire manufacturing facility is an iterative process. The calculation of loadable hours is repeated for each work center, and the results will become part of the permanent files for planning future capacity requirements.

To cite an example, retrieve the 1,000 pieces previously used to construct the product load profile. This standard load takes 11 hours to set up and run. Assuming that it will run on the equipment for which the measurement formula computed 129.6 (rounded to 130) standard hours of load, these 11 hours would be plotted against all other orders, both planned and released, competing for the same 130 hours within the constraint of the due date on the order. If simulation becomes necessary to level the load, a more or less defined priority sequence can then be established.

Capacity requirements planning here is in reality merely a forecast. A forecast assumes, excepting any known extrinsic factors, that the future will resemble the past and that there is no more definite tool available. This is certainly the case here. Bearing this in mind—whether you are using load profiles in the rough-cut planning process to verify the MPS or the capacity measurement technique to prioritize the MRP—it must be reemphasized that these are estimates only. The only other generally recognized technique is to gut a chicken by the light of the full moon and note the direction in which the entrails fall. Then you'd know for sure.

## FINITE CAPACITY CONSIDERATIONS

At this level the MRP output—planned orders and CRP output—planned hours come together to be acted upon, and this is where planned orders become released orders.

### Machine Loading

There is a certain amount of control that must be exercised at the point of input. To issue shop orders to the various starting operations merely because they become mature, irrespective of where they fall in the scheme of things, is certain to generate confusion and more and more work-in-process, which, in turn, will cause more and more expediting. The greater the number of orders in the shop that are not being (and cannot be) worked on, the more this will eventually negatively impact customer service. That is why the rule should be to put no more work into the shop than the shop can work on that day. (This goes for succeeding operations as well.) If one day is not feasible, then certainly not more than half a week.

This is probably the most difficult of all tasks. Just as it is human nature to try to squeeze a bit more into the MPS, people try to squeeze a little more out of existing capacity. Schedule changes are going to disrupt planned priorities—that, you have no control over. However, input control of releasing orders to the shop is the only way to get a handle on what *can* be done versus what *must* be done. It is also the most logical routine for keeping work-in-process within reasonable limits.

### Operations Sequencing

This is the final step in the planning process. Operations sequencing is a type of finite loading, or loading the machines to known and available capacity. It employs forward-scheduling as its functional activity. Thus, it is easy to tell if due dates are going to be realized by totaling run and lead times of the required operations. In addition, in a computerized facility, this will automatically reschedule backlogs at bottleneck operations when overloads occur. The literature on this subject often lists five common techniques for priority sequencing through all the various operations. They are first in, first out (FIFO); last in, first out (LIFO); shortest processing time in which orders that run first are those with least time remaining for completion; remaining slack time, or how long to due date versus how much process time left; and critical ratio. It does not seem at all unusual in this age of buzzwords that what all such lists fail to include is the best sequencing technique of all: common sense.

## CAPACITIY CONTROL

### Input/Output Controls

No manufacturer controls output. It is the other way around. To be sure, certain features can be included in your system to adjust some of the characteristics of output. But either output is going to happen or it's not. You may be able to help it along, but you certainly are not going to control it.

This is what is so ticklish about scheduling downstream operations. You cannot feasibly schedule a succeeding operation until you know the output of the preceding operation. Despite the foregoing, the term *input/output controls* is not really a misnomer. The control feature lies in what you do with the numbers once you have them. What kind of strategic planning can take place? What decisions can be made once you are working with hard-and-fast information?

The input/output report is the central feature of the capacity control procedure. It analyzes and compares the actual input versus the planned input and the actual output versus the planned output. The deviation of actual versus planned as calculated from periodic reports (preferably daily) becomes the basic data upon which subsequent decisions are based.

The planned input numbers are an outgrowth of MRP/CRP routines. MRP determines the quantity and timing, and CRP figures the standard hours of load within the time constraint. Planned output, identical to the MPS, is what is expected to be obtained for the given period. Any deviation that is derived by comparing this plan to the actual will directly affect the backlog. It is through the use of these numbers, therefore, that the backlog (the amount in queue) can be controlled. Figure 7–1 exhibits the features of a typical input/output report.

**FIGURE 7–1**
**Input/Output Report**

| | | Week | | | | |
|---|---|---|---|---|---|---|
| | | 1 | 2 | 3 | 4 | 5 |
| Planned input | | 15 | 15 | 0 | 10 | 10 |
| Actual input | | 14 | 13 | 5 | 9 | 17 |
| Cumulative deviation | | –1 | –3 | +2 | +1 | +8 |
| Planned output | | 11 | 11 | 11 | 11 | 11 |
| Actual output | | 8 | 10 | 9 | 11 | 9 |
| Cumulative deviation | | –3 | –4 | –6 | –6 | –8 |
| Actual backlog | 20 | 26 | 29 | 25 | 23 | 31 |

Source: T. E. Vollman, W. L. Berry, and D. C. Whybark;, *Manufacturing Planning and Control Systems* (Homewood: Dow-Jones Irwin), p. 134.

## The Critical Ratio

The final checking feature to provide a measure of ongoing control is the critical ratio, which is a neat piece of work. Born of the desire to control inventories—back in the days of scientific inventory management, when P&IC people were all training to become production and inventory "engineers," this was another in a growing list of formulas that proved beyond any doubt that they were truly engaged in a scientific endeavor.

So much for the good old days, when so many of these seemingly logical techniques failed miserably in actual practice. In this instance the underlying idea was to review the stock position of an item on a regular basis and calculate its priority based upon its quantity on hand relative to its order point. What the theory was utterly unable to deal with was that while many orders would be in a priority position, only a few would show current requirements outstanding against them.

This was in the days before MRP, of course. The current practice is to run a critical ratio for every part in the open-order file, irrespective of whether it has begun its journey through the shop. This seems to be redundant, for this could just as easily be handled through the normal MRP rescheduling process.

Over the years, the critical ratio has undergone some transformation as it has been applied to a variety of situations. Originally it was a product of ratios A and B. Ratio A was the percentage of stock depletion as withdrawals from stock brought the part count closer to its order point. Ratio B was the percentage of work completion as the replenishment order was being processed to coincide with the due date. The critical ratio was the measure of these two. The lower the value of the ratio, the higher the priority of the job, and vice versa. Not all that long ago, when the big craze was setting priority rules, this approach was termed the *dynamic priority approach rule.*

Recently interest has centered on a newer type of critical ratio, one based on due dates. Under this scenario today's date is subtracted from the due date and the difference is divided by the cumulative lead time (or lead time remaining if processing has begun). A low ratio suggests a high priority, a ratio of 1.00

signifies an order precisely on schedule, and a ratio greater than 1.00 means that the order enjoys some slack time. This is similar to the original formula, but all relationship to inventory has been removed. It is simply a ratio of work remaining to time remaining. Normally the job with the lowest critical ratio will run next.

Whether or not the critical ratio is employed in any of the above pursuits (except to control inventories) is a matter of personal choice. It all depends upon what works for you. However, its use under capacity control and the input/output (I/O) report will perform a function different from those previously mentioned. In this case the critical ratio is a technique that calculates the current status of a job in process. To put it more pedantically, it is a technique for the dynamic updating of operation priorities. It can be expressed either as a ratio of time remaining to work remaining or through a cumulative updating procedure. One of the primary differences between this use and those previously mentioned concerns lead time. Capacity determination in either the planning or the control stage would be hard-pressed to handle a concept such as cumulative lead time. As a result, the I/O report is only concerned with the lead time of a single machine or work center; hence the term *operations sequencing*. Yet this actually works out to be an advantage. As the parts proceed through an operation, the I/O report will state on an ongoing basis the exact status of a requirement. It is really a matter of indifference how much of the total lead time has been consumed; the concern is only for *a part* running on *an operation*.

If an order quantity of 500 were to be run over five days, cumulative updating would be shown on the I/O report as one total production figure as the days are consumed. It could be presented as a percentage of 100 or a decimal equivalent for the first day. On the second day the denominator would be 200, and the report would show a cumulative total produced for days one and two; and so on throughout the week. Those operations whose calculated value are on the low side could be singled out in an exception report for individual handling. The following chart shows the kind of feedback information a report like this would provide:

| | *Scheduled* | *Actual* | *Ratio* |
|---|---|---|---|
| Day 1 | 100 | 90 | .90 |
| Day 2 | 200 | 170 | .85 |
| Day 3 | 300 | 250 | .83 |
| Day 4 | 400 | 290 | .73 |
| Day 5 | 500 | – | – |

The critical ratio reported in this fashion is in effect a form of tracking signal and is used to compliment the I/O report. The report provides feedback on a weekly basis, while the critical ratio handles the individual days. Figure 7–1 reveals an operation slipping precariously into arrears. An exception report should materialize, if not by day 3, certainly by day 4.

Some practitioners believe that an accurate input/output report obviates the need for the critical ratio. This is all right if you are not bothered by expensive surprises. In the case of the 500 ordered in the example above, if you are only going to get production of something less than 400 during five days, it would seem likely that you would want to know that before it happens.

## SUMMARY

This discussion of capacity considerations was not meant to be all-inclusive—nor was the discussion of material requirements planning that preceded it. The purpose here was simple and straightforward. Over the years, the various concepts and terms, through alterations and additions, have taken on somewhat different meanings and their original intent has become obscured. Our goal here has been to explain in understandable form those principles and practices that focus on serving and sustaining customers. Your survival depends on it.

The discussion of capacity has not been complete because some aspects of capacity better lend themselves to Just-In-Time practices, which have not been included here.

Many MRP systems have not been utilized to their maximum potential. Too often the advertised primary benefits have centered on inventory considerations, at the expense of other manageable assets, capacity being a likely candidate. Unless the necessary planning, execution, and control take place throughout the system as a whole, the potential advantages will not be realized and the effectiveness of the manufacturer will be diminished.

# PART 4

# JUST-IN-TIME MANUFACTURING

# CHAPTER 8

# THE PLAN

The Just-In-Time (JIT) philosophy and its approach to productivity and efficiency is the very embodiment of common sense manufacturing. The two terms are very nearly synonymous. Yet in order to better understand what is involved here, it is necessary to dissect and analyze the various principles and technology that have evolved over the past 50 or so years of manufacturing activity.

## A NEW WAY OF ADDRESSING PROBLEMS

Most experts agree that you have to acquire a whole new way of addressing problems; you need to develop a new mindset. To put it another way, you have to approach this with a closed mind, as it were. What you are setting out to do is going to work, is reasonable and logical, and has helped company after company reach new heights of efficiency and profitability and will accomplish the same things for you.

In a list of things that must be done, probably the most important is shedding the trade-off mentality. No longer can you say, "Sure, I can give you better quality, but it's going to cost you" or "I can give you a lower price, but quality is going to suffer." Now the only permissible thing to say is: "I can supply your product to you with your level of quality and at a competitive price, and this is how I'm going to do it." And do it. This is what George Orwell would call "new think."

## JAPAN'S SUCCESS

Some individuals who have done considerable research on Japan and Just-In-Time have concluded that much of their success is attributable to their unique culture, their particular commercial practices, and the laws governing them. Wholesale adoption of their practices, they say, would be impossible—and probably illegal. Much to their credit, however, the Japanese show a healthy respect for the fragile underpinnings of a free-enterprise system and therefore are not shackled with antitrust laws and other forms of punitive legislation.

Regarding Japan's labor history, many of their routine practices are anathema to American trade unionists, particularly their long-standing position regarding work rules, job classifications, and seniority.

Japan's geography is another consideration. Relatively speaking, nearly everything in Japan is close to everything else, and it's a simple matter to ship critical components once—even twice—a day from one side of Tokyo to the other. This is hardly comparable to shipping daily between Indianapolis and Grand Rapids and back again.

There was no master plan designed by forward-thinking leaders in Japan 30 years ago to get them where they are today. Much of their success can be attributed more to accident than to design. In fact, the Japanese have never been known as great innovators, but rather as very astute imitators. It was this that led them to bring to their shores the likes of W. Edwards Deming and Joseph M. Juran, among others, who were commissioned to put Japan's economy on the track that would build an industrial base like the one we had in the United States. Of course they first had to rid themselves of their terrible image of manufacturing low-quality products. To do this required coming here and recruiting the most renowned names in the field of quality control. And, as the saying goes, the rest is history.

One accident that worked in Japan's favor concerned their treatment of inventory. At the end of World War II Japan was broke. Their debilitated industries had no money to lay in huge stores of material. Even if they had had money, space was at such a premium that there would have been no economical way to carry them. So they made do with what they could afford to process at one time. As they became increasingly prosperous, they saw no reason to unlearn the lesson that was born out of necessity. In due time they got the recognition they well deserved for this accident of history.

Japan does not hold a patent on just-in-time. Many of the concepts that comprise this philosophy were originally introduced in Henry Ford's auto assembly plants back in the 1920s. It was the Japanese, however, who implemented them on a grand scale and became remarkably successful in doing so. If one were to search for a cause of this apparent turn of events, it might be found in the decades of the Depression and the postwar economy, when concepts such as these would appear to have been extreme measures, without justification. Now, however, they are entirely justified, and have been for some time.

## A NEW WAY OF SOLVING PROBLEMS

American industry, by and large, approaches problems by employing one or more of three alternatives: It throws money at them; it throws people at them; and/or it throws computer programs at them. Rarely do these techniques accomplish anything but mask the problems so that they can be forgotten for a time. The Japanese, on the other hand, have come up with an exceptionally sound problem-solving technique called the "five whys." It says that if you put the question Why to a problem five times, by the fifth answer you will have uncovered the root cause. That is the problem you solve.

The Japanese five-whys system of problem solving is certainly superior to the American "one who" approach, which says, "Find the one who is responsible and fire him." This is somehow supposed to bring order to the system. All it actually accomplishes is to transfer an experienced employee to the competition because he has proved that he is human. Furthermore, the "one who" doesn't even have to be responsible for the problem, just visible and available.

Probably the most appallingly senseless and totally counterproductive practice employed by American businesses is what everyone knows as the c.y.a. syndrome, which posits that the single most productive endeavor one can engage in is to ensure that his or her "*a* is always *c*'ed." For example, early in this writer's career, while working in the home office of a large multinational corporation, a crisis fell out of the blue one day. Precisely what precipitated it is lost in antiquity, but the furor it caused was memorable: a bunch of highly paid executives rushing to their filing cabinets and feverishly pawing through their records to make sure they could not be held accountable for dropping the ball. The really troubling part is that this has become a corporate ritual that is continually played out. And how many problems has it ever solved?

## DEFINING JUST-IN-TIME

Defining Just-In-Time is not an easy task. It is perceived differently by different people because it is intended to accomplish so many different things. One might say that it is a system of adding value to a product through the systematic elimination of waste, but such a definition is too restrictive. It is the purpose of the JIT strategy to pervade the entire manufacturing environment, not just to eliminate waste, although that is its primary function.

Another definition might be to give the customer the right product, in the right quantity, at the right place, at the right time. Certainly this customer focus is at the heart of JIT principles, but it is more a result than an application—what you hope to achieve through efficient resource utilization.

Therefore, it is easier to describe Just-In-Time by examining its more common characteristics. First, however, it is important to ask whether they are truly new, genuinely revolutionary as advertised, or merely a collection of old systems and techniques packaged differently. In effect, JIT is both. Certainly no knowledgeable person is going to say preventive maintenance or housekeeping are new. But they take on a new significance within the strategy of JIT, and it does not take a great deal of imagination to see how they fit neatly into the value-added/waste-elimination concept.

Entirely new, however, are principles like flexible manufacturing, product and process flow, and the rather novel idea that improved throughput is a far more accurate gauge of productivity and efficiency than machine speed and pieces produced per hour. The "pull" system likewise falls into this category.

Quality is not a new concept, but the way of achieving it is. It is no longer acceptable to run a large number of parts and hope they all meet spec. You must build quality in at the start of the run and monitor it throughout. This is probably the single most important aspect of JIT, and unless it is addressed early and completely, your project is doomed.

Level-loading is another current tactic that has never been solved satisfactorily by conventional means. Smaller runs and quicker setups (i.e., improved throughput) are a little different approach to getting there.

These are just a few of the concepts that make up Just-In-Time manufacturing. It's interesting to note the common thread in each: They are all basic common sense.

## BENEFITS OF JUST-IN-TIME

It would be useful at this point to list the advantages that accrue from implementing JIT manufacturing. If you are going to embark on a course that will change forever the way you operate your business, you will want to know beforehand what you can expect as a result of all this effort.

1. First, obviously, there must be some benefit to the customer. If improvement in customer service were not realized, JIT would merely be an exercise.

   Although customer service is always the primary consideration, there are many other benefits, listed here in no particular order.
2. Reduced inventories and an improved inventory turnover.
3. Improved employee morale.
4. Improved productivity.

5. Better customer-vendor relationships.
6. Reduction in floor space through better plant layout and gradual elimination of work-in-process.
7. Reduced labor in stockroom/stockkeeping personnel.
8. Reduced record-keeping.
9. Higher return on investment.
10. Eliminating the "end-of-the-month crunch."
11. Improved quality causing less scrap and rework.
12. Reduction in manufacturing lead times.
13. Fewer schedule changes.
14. Elimination of all trade-offs from planning and execution.

These are the major benefits, but the list could go on and on. As one can see, each of these benefits accrues as a result of some definitive action. Yet the objective of each action taken is not primarily to realize one of the above-listed items, but rather ultimately to eliminate waste and/or to more efficiently utilize limited resources.

If one holds to the current belief that during its life in the manufacturing cycle a part is actively being processed only about 10 percent of the time, then obviously 90 percent of the time it is in some other, inactive state where no value is being added. And when no value is being added, waste is occurring. These are the only alternatives. The inactivities that make up the 90 percent become the prime candidate for scrutiny.

Two aspects of this new way of thinking come to bear here and are part and parcel of any action taken: predictability and visibility. You know what is going to happen, and when it does, you already have a plan prepared.

A brief example will illustrate this. You know from experience that each time you run a particular part, the die needs to be sharpened after about 2,000 hits. You receive an order for 2,000 parts. It is standard practice in your shop to overrun orders that are running well because efficiency is one means by which performance is measured. But how much sense does it make to overrun this part and accumulate 3,000 or even 4,000 parts, all in the name of efficiency, when the possibility exists that you will end up reworking 30–40 percent of the run? And yet the system seems to encourage this.

How much better it would be to shut down the machine after 2,000 strokes and be able to predict precisely when this will happen. Even if the order were greater than 2,000, you shut the machine down at this point and resharpen the die automatically. The operator and crew can engage in some alternate activity because you have planned for it.

## JUST-IN-TIME PREREQUISITES

In order to put JIT practices into operation, a construction blueprint, or a step-by-step progression, is necessary to guide the way over an extended period of gradual improvement toward your objective.

It is difficult to cite a typical example of Just-In-Time implementation because many companies employing this strategy are atypical. Certainly JIT suggests repetitive manufacturing, but this doesn't exclude firms that are not similarly predisposed. It is important to remember that JIT is a mental process as much as a physical one and that it is more a commonality of purpose than a technological breakthrough when two disparate manufacturing types arrive at the same stated objective. After all, the list of advantages does not arise from one type of technology to the exclusion of all others. It is broad enough to apply across the board. It is the getting there that is important; just how it is done is not the primary concern.

In the following sections we will focus our attention on the objectives of JIT. But first we must lay a firm foundation, comprised of three inescapable modules: MRP, MPS, and quality. Each will be examined in turn.

### Material Requirements Planning

It has been stated in a number of magazine articles that a material requirements planning system is as important under a JIT approach as it is with conventional manufacturing. Furthermore, it is commonly believed that when a firm makes the transition into JIT, the MRP system must be altered in such a manner that it will not run counter to what you intend to accomplish under JIT. It should be reiterated: This is not necessary under the rules of common sense manufacturing, as outlined in these pages. MRP needs no revisions to move in the direction of Just-In-Time. This is one of the things that will occur naturally under a common sense approach.

The only major difference any practitioner will experience is in the execution phase. It goes without saying that with any manufacturing strategy requirements must be planned and scheduled. Under JIT the dissimilarity will be noticed in the quantity and timing of individual receipts. What you hope to achieve in the end is to bring *into* your shop each day precisely the amount that you will process that day and *through* your internal shop only the amount that you will ship that day. Figure 8–1 shows how gross requirements change from a weekly need to a daily production quantity.

How you accomplish this within a framework of a system that embraces thousands of active part numbers—not to mention the logistical nightmare of scheduling the receipts from countless vendors—is one of the problems facing anyone who wishes to tackle JIT manufacturing. These problems will be discussed further as we go along, but you should never lose sight of the bedrock principle: gradual improvement.

**FIGURE 8–1**
**Gross Requirements under Just-in-Time**

| | Week | | | | | | | |
|---|---|---|---|---|---|---|---|---|
| | 14 | 15 | 16 | 17 | 18 | 19 | 20 | 21 |
| Gross Requirements | | 100 | | 100 | | 100 | | |

| | Week 15 | | | | |
|---|---|---|---|---|---|
| | M | T | W | Th | F |
| Gross Requirements | 10 | 10 | 10 | 10 | 10 |

## Master Production Schedule

Like MRP, the master production schedule (MPS) loses none of its importance along the road to Just-In-Time. As a matter of fact, it promotes an idea that is fundamental in its makeup and operation: the concept of load-leveling. The entire operation of a JIT system revolves around this principle.

It was recognized under the list of advantages that first and foremost improved customer service was the primary thrust of all activity. The objective is to satisfy customer demands as needed rather than ship a truckload from the finished goods inventory and then replenish the supply with a big slug from production, much like what is being done right now with only varying degrees of success.

The operational characteristics of a load-leveled MPS parallel very closely those that activate the MRP system, that is, more frequent scheduling of ever smaller amounts. If your manufacturing environment calls for the use of a final assembly schedule, this will become the controlling mechanism. Scheduling the amount you need from the back end causes the front end to become scheduled automatically. This quantity moves through each operation in the shop or down the assembly line in the same leveled amount.

The benefits of a load-leveled MPS can be seen in the examples shown in Figures 8–2 and 8–3. It is assumed that among the products of a given firm are three finished items, labeled A, B, and C, respectively. The assembly schedule for these three products is as follows:

**FIGURE 8–2**
**Load Leveling under Just-in-Time: Before**

| *Week 1* | *Week 2* | *Week 3* |
|---|---|---|
| *Product A* | *Product B* | *Product C* |
| Assemble 100 Units | Assemble 100 Units | Assemble 100 Units |

**FIGURE 8–3**
**Load Leveling under Just-in-Time: After**

| | *Week 1* | *Week 2* | *Week 3* |
|---|---|---|---|
| | *Product A* | *Product A* | *Product A* |
| | Assemble 33 Units | Assemble 33 Units | Assemble 34 Units |
| | Product B | Product B | Product B |
| | Assemble 33 Units | Assemble 34 Units | Assemble 33 Units |
| | Product C | Product C | Product C |
| | Assemble 34 Units | Assemble 33 Units | Assemble 33 Units |
| Total | 100 Units | 100 Units | 100 Units |

The logic for this conventional type of scheduling assumes that it is the most cost-effective and efficient way of satisfying the majority of customer orders. However, this approach is company-focused, not customer-focused.

Assume it is Monday of Week 1 and one of your better customers has been promised 25 units of Product C on Thursday, but your inventory position is somewhat less than that. This type of problem occurs routinely in all manufacturing environments. There are just two ways to resolve it: You can tell the customer that Product C won't be available for two more weeks—which no one in his right mind would even consider—or you can change the schedule to meet the demand, with all the disruptions this causes up and down the supply chain.

Just-In-Time proposes a solution to this type of problem that is simple and logical, yet at the same time maintains assembly output at 100 units per week. Figure 8–3 shows how a load-leveled schedule would be modified.

How one arrives at this position is precisely what Just-In-Time is all about. Once arrived, it then becomes the starting point for even further improvements. JIT is a continuing, never ending series of incremental steps toward excellence. Like work, it is never done.

## Quality

It would be difficult indeed to envision Just-In-Time without the functions of quality control first being in place. The system cannot operate smoothly unless the constant interruptions of out-of-spec parts are brought in control. (The key phrase here is in control.) No longer is it permissible to run a mass of parts and then inspect them to see how many bad ones there are. This is one of the most expensive forms of waste. Not only have you wasted resources by running defective parts, but you must waste them again by rerunning or reworking the shortfall.

The only reasonable way to bring an abrupt halt to this problem is to institute quality control at the source, and this can only be accomplished through statistical process control (SPC). Of all the statistical techniques that have been introduced to plant operations, none has been so effective. This is because it works.

In its usual form, SPC is based on the assumption that the variations that occur among individual pieces within a run, or between lots in successive runs, are attributable to either special causes or common causes. Special (assignable) causes are those that can be corrected by line people as they occur. They are unpredictable and must be remedied by the operator, setup man, shift supervisor, applications engineer, or supplier technician.

Common causes, on the other hand, are those characteristic of a given process, experienced by all users. Common causes can only be corrected by management intervention (e.g., retooling by engineering). The objective is to eliminate or at least dampen the effect of both of these sources of variation so that the process will be under (or in) control. A process is said to be under control when random samples pulled during a production run exhibit a constant pattern of variation.

To be in control, however, has nothing to do with parts being within specifications. It means that you have achieved a pattern of variation that is stable and predictable. It is the first step toward achieving process capability—the ability to run parts that are within specification limits.

To determine whether a process is capable, one must first determine that it is in control. The form for measuring progress is the variables control chart, generically called an $\overline{X}$ (X bar) and R chart (see Figure 8–4). $\overline{X}$ is the universal symbol for the mean, or average, of the sample of the population being studied. Under SPC it means the average of, say, five successive pieces pulled during a production run at stated intervals for determining the properties of a given attribute (size, weight, shape, etc.). The average ($\overline{X}$) of the sample is plotted on the $\overline{X}$ portion of the control chart, and, as successive $\overline{X}$s are calculated, an $\overline{\overline{X}}$ (X double bar) will begin to emerge. An $\overline{\overline{X}}$ is called "the mean of the means." It will become the centering location for all plotted sample averages. Dispersed above and below this line in a (hopefully) random fashion will be the individually plotted sample means, or $\overline{X}$s.

The R in $\overline{X}$ and R stands for *range*. It is important to know not only the average value of a given sample but the range within the sample of values from highest to lowest. For example, in a sample of five pieces, the pieces range in value from 5.5 to 4.5. The $\overline{X}$ would be somewhere between these two values, but the R would be 1.0. Each R is plotted on the chart whose graph will be below that of the X. Like the mean of the means, the mean of the ranges ($\overline{R}$) can be calculated.

Thus, the X is for centering or location and the R is to measure variability or spread. In order to determine whether a process is in or out of control, limits for the value of the characteristic in question must be established. These are calculated according to a specific formula, and the numbers thus computed will show the upper and lower control limits of both the X and R values. When any sample mean or spread plotted on the chart falls outside these control limits, the process is out of control and should be shut down.

While the exact interpretations of statistical process control and the formulas involved in constructing trial control limits are outside the scope of this book, they can be gleaned from any text dealing with these subjects. However, it is important that anyone interested in Just-In-Time become familiar with the concepts involved. Monitoring a process using the principles of SPC will tell you not only how high is the quality of the output but when a process is tending to go out of control, allowing for necessary corrections and keeping scrap and rework to absolute minimums. This is truly achieving quality at the source. That this procedure is so far superior to the conventional practice of 100 percent inspection goes without saying. You are attempting to work smarter, not harder.

A sample variables control chart is shown in Figure 8–4.

## Mechanics of Just-In-Time

Perhaps the best way of introducing the mechanical side of Just-In-Time is to look at how it would change what is called conventional manufacturing. How would current manufacturing standards and practices be altered by stiffer competition and force us to adopt a JIT recipe?

In an earlier chapter, a typical order was entered and followed through a typical manufacturing plant. It needs to be recalled here to illustrate a different principle. The typical order arrives through order entry, or a forecast quantity comes due. With copies to all requisite departments and with a bill of material defined, it finds its way onto the master production schedule. When its time comes, it is scheduled through the shop in some sort of logical sequence. After making its way through the system, the order is final-finished or assembled, packed, and sent to a holding area for shipment to the customer according to his due date.

This is the way it is *supposed* to work, and it does, but in a somewhat more discontinuous fashion. Material shortages, schedule changes, machine breakdowns, and 101 other routine glitches all contribute to the enduring problem of serving the customer.

This form of manufacturing is termed *batch processing,* which means that the particular part being processed is one of a quantity, or batch, of the same part. It is held at the work station until the entire batch is completed; the entire batch is then moved as a group onto the next work center, where some other batch is probably being worked on. When its turn comes, this group of parts will again be processed as a batch, and this activity will be repeated continuously as orders enter and leave the production areas. When it reaches the final stage, it will again be finished and packed as a batch.

FIGURE 8–4 Variables Control ($\bar{X}$ and R)

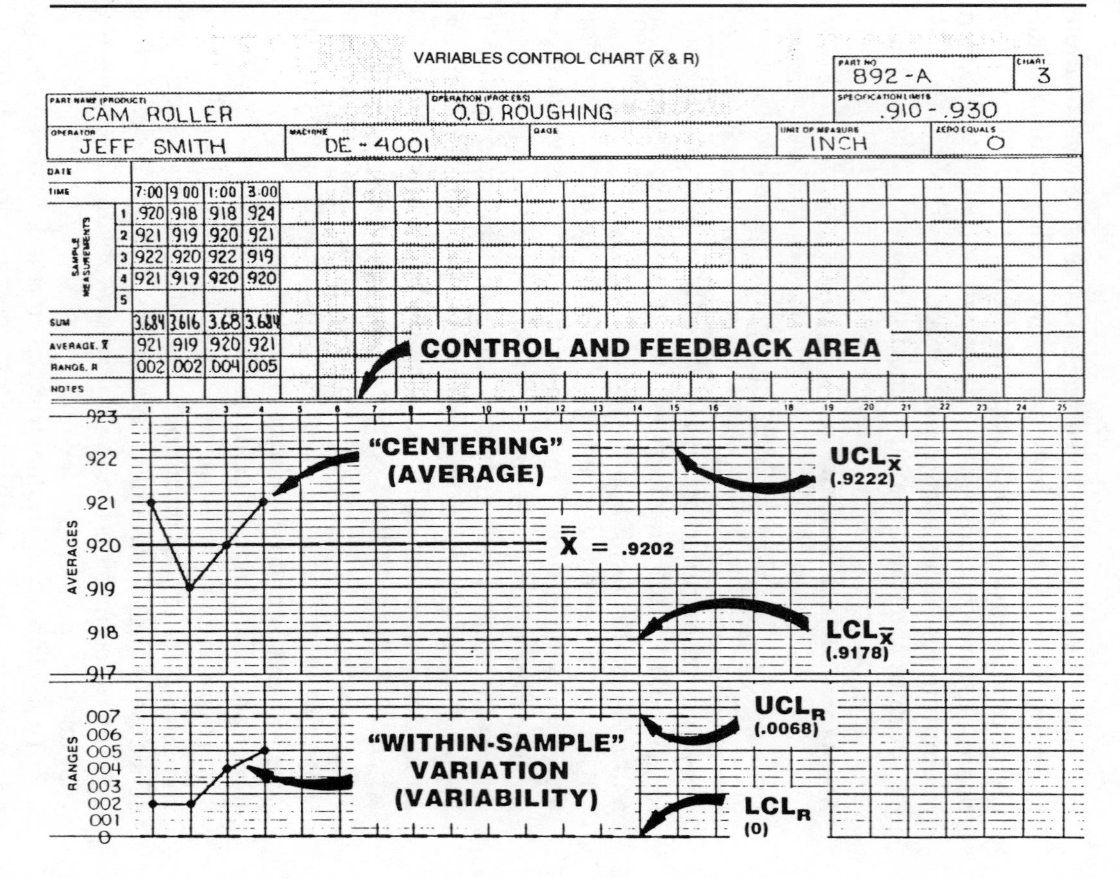

This process can be visualized by the use of a few simple exercises. Assume that an order for 100 pieces enters production through the normal channels. Further assume the order will pass through four stages of fabrication, at which point it will become an assembly component of the finished product. A review of production records suggests an allowance of 10 percent for spoilage at each of the four processing points.

At the beginning of this procedure, enough material is issued to make 100 pieces, and the part enters its initial operation.

| | | |
|---|---|---|
| | *1st Stage* | |
| 100<br>×.10<br>10 | processed<br>spoilage | Ten pieces are spoiled, leaving 90 to go to station 2 |
| | *2nd Stage* | |
| 90<br>×.10<br>9 | processed<br>spoilage | The process is repeated at this station, leaving 81 to go to the third work station |
| | *3rd Stage* | |
| 81<br>×.10<br>8 | processed<br>spoilage | This leaves 73 pieces to proceed to the final stage |
| | *4th Stage* | |
| 73<br>×.10<br>7 | processed<br>spoilage | From the original order of 100, only 66 units can be assembled |

What about the remaining 34 pieces? Is it feasible to take the time to set up and run only 34 pieces, with all the changing around that requires? Of course the answer is no. The obvious solution to the problem is to add the 10 + 9 + 8 + 7 pieces, or 34 pieces onto the original order before the first operation. The order for 100 pieces will then become one for 134.[1]

[1]The author is fully cognizant of the fact that a 10% spoilage factor worked back from 134 yields only 88 parts; and working forward would require a starting quantity of 152 to get

So each time this particular component is run, 134 will be started in order to feel relatively certain that 100 end items can be assembled. If each piece cost $5, you would have to eat or incorporate somewhere in the selling price $170—and this is just one of many parts that goes to make up the end item.

Suppose, for the sake of argument, that spoilage could be cut in half by doing something slightly different, maybe even a minor adjustment. Without having to go through the computation again, the spoilage works out to about 19 percent. In other words, the increase in productivity is 15 percent, and it didn't cost anything. This is what the elimination of waste, the value-adding concept is all about.[2]

This points to yet another principle that is basic to JIT. A one-word question says it all: Why? Why be satisfied with a spoilage factor, not of 10 percent, but of a whopping 34 percent? If you put into the system enough material for 100 pieces, you should get 100 pieces out the other end—nothing less. If you do not, something is wrong somewhere in the system.

Look at the same problem from another perspective. Assume that examination of recent records for runs of Component X reveal that spoilage does average 10 percent per operation, or 34 percent over its entire run. However, digging a little deeper reveals that, while the total spoilage may be about 34 percent, the individual percentages making up this total vary greatly among themselves. Suppose the spoilage breakdown looks something like this:

---

100 parts. However, concentrating on the mathematical precision entirely misses the point being made here. One hundred thirty-four is a number that is reasonable and practical from a manufacturing point of view. It was chosen for purposes of illustration only.

[2]Computation showing the effect of a 5 percent spoilage under batch processing:

$$100 - (.05 \times 100) = 95$$
$$95 - (.05 \times 95) = 90.25$$
$$90.25 - (.05 \times 90.25) = 85.75$$
$$85.75 - (.05 \times 85.75) = 81.50$$

These decimals are not precise, but they are close enough. Shown are average spoilage rates of 5 percent, 4.75 percent, 4.5 percent, and 4.25 percent, totalling 18.5 percent for the entire manufacturing cycle.

| | |
|---|---|
| Operation 1 | 4% |
| Operation 2 | 17 |
| Operation 3 | 9 |
| Operation 4 | 4 |
| Total | 34% |

This kind of mini-Pareto breakdown is extremely revealing. It certainly highlights the area to which activity should be directed in order to yield more immediate returns. One fact remains undeniable: Continuing to operate under these conditions causes standards to become permanent fixtures. Unless management puts forth the effort to improve the system, it is likely never to change—except to get worse.

The JIT principle of gradual improvement is clearly evident here. For every decrease in shrinkage that can be accomplished, there is a corresponding increase in productivity. Furthermore, it often involves little or no cost. Each time something is input into the system expected output must be 100 percent. That is your goal. Maybe this goal is not attainable but you won't know if you don't try.

Conventional manufacturing, typified by batch processing, provides a perfect illustration of a push system. A push system is one where parts are loaded at the beginning and pushed through the factory. Just-In-Time is characterized by a pull-type mechanism, where the focal point is at the opposite end of the shop. Here the operation in final assembly pulls the components from the preceding station based upon need and rate of assembly. Due to its uniqueness, the pull system is usually one of the last to be implemented. It requires a series of steps that must first be in place in order for it to be effective. This is treated more fully later in the next chapter.

## CATEGORIES OF WASTE

Earlier in this chapter, when we were searching for a definition of JIT, one of those offered was "a system of adding value through the systematic elimination of waste." This definition is correct, only as far as it goes and so it is necessary to identify where waste can exist in a manufacturing facility. There are six categories of waste, the first being by far the most important:

1. Inventory: By its very nature, inventory is a waste. That it is the basic life support system for any firm engaged in manufacturing does not diminish this fact. In fact, it serves to magnify it.

   Some inventory control texts and articles go into great detail concerning the functions and classes of inventory. Rather than get bogged down in unnecessary verbiage, only the three most common (and therefore, most wasteful) types of inventory will be mentioned:

   Fluctuation inventory, a.k.a. safety stock.

   Anticipation inventory: production in response to a forecast of demand.

   Work-in-process inventory: the most insidious form of inventory and the most difficult to control.

   As JIT strategies are discussed, the way to limit the effects of inventory will be evident. This applies to all except anticipation inventory. In order to maintain a level operational load, a portion may have to be made in advance. This is where a creative marketing department can come up with a list of incentives.
2. Waste involved in waiting: This is our old friend queue time. It has been stated that within the manufacturing cycle, a part spends only 10 percent of the time actually being processed. Of the remaining 90 percent, the preponderance of this is wait (or queue) time, thus becoming a prime candidate for waste reduction.
3. Waste in making defective products: This is actually a double waste. First, additional resources must be consumed in replacing defective parts; and then defective parts must be reworked.
4. Waste of superfluity: Implied within the concept of added value is supplying only what the customer ordered and is willing to pay for. Overspecifying details or running an excess of parts is a waste of resources that

could be used otherwise. Recall the second definition: "the right quantity of the right part, etc."

5. Waste in motion: Anytime production parts are put into and then removed from a stockroom or storage facility, coupled with a transaction recording in and out, it is a total waste. No value is being added, but cost is. The same goes for costs incurred in inspection, whether in-house or vendor-supplied parts.
6. Waste in transportation: This is a two-part waste reduction problem. The first is movement inside the plant from one processing point to another; it can be solved by a more efficient plant layout. The second part is movement outside the plant. For example, if inbound freight charges are a flat rate, whether for a 1-day or a 10-day supply, extra freight costs vis-à-vis carrying additional inventory, obviously, cannot be discounted. Due to all the external problems involved, this is more a traffic problem, which puts it outside the bounds of this discussion.

# CHAPTER 9

# THE ACTION

A company subscribing to the ideas of just-in-time will by now have a pretty good perception of what it is they have involved themselves in. A successful implementation of JIT will depend largely upon the foundation that has been laid. While there is no prescribed procedural path to follow, there are several general principles that can serve as a guide.

The first is a plan for action: the procedural steps the firm decides it will follow. Another is the selection of a pilot operation on which tests can be performed. Learning your mistakes is important in the early stages of implementation. The third principle is employee training and education, which will continue throughout the implementation phase and beyond. Training sessions will be attended by three main groups: those involved in the changes (accounting, engineering, etc.), those affected by the changes, and those responsible for the changes. The last group is presented with a unique challenge, for management interest and involvement is nowhere more important than in their direct participation in this activity.

Because there are no hard-and-fast rules for implementing JIT, the following discussion of various concepts is selective and not presented in any prescribed order.

## FLEXIBLE CHANGEOVER

Throughout the history of American industry, setup and equipment changeover has always been considered a given. Its cost is said to be included in the price of the product, and we all have been content to live with it. The common scenario in which a changeover takes place is something like the following.

A machine shuts down after running the last part on an order. The die is removed and toted back to the tool room. The operator inquires about the status of the next die and is told that it is nearly done—"about 15 minutes"—so the operator goes on break. After several cups of coffee in the next 20 minutes, he meanders back to the tool room and is told that the die is ready. Great! First he must scout up a die cart. After a while he wheels in a cart, loads up the die, and returns to the machine to begin setting up. Upon reading the instruction sheets, he remembers, belatedly, that this die requires special calibration tools and trudges back to the tool crib to acquire them. If no one is in front of him, he secures these and proceeds back to his machine to continue with the changeover.

How much time has now gone by? An hour? An hour and a half? Admittedly this was a little exaggerated, but it does make a point. The biggest problem you have under these conditions is that a mindless changeover, wasting valuable time, becomes a self-perpetuating practice. "What's the big deal? This is the way we've always done it."

Suppose instead that you had the newly sharpened die sitting on a die cart, with the special tooling close at hand and the instructions from the setup manual fresh in your mind, just as the machine completes its final stroke. This is what rapid changeover is all about, and it has not cost you a blessed cent. You will have worked smarter, not harder.

What needs to be done when analyzing a setup procedure is to separate external and internal time. External time is the time involved when the machine is running its current job. Internal time is the time when it is shut down for changeover.

There is an optional step that can be taken here that may play a very important part in the analysis: videotaping each setup procedure, preferably with a stopwatch or digital clock running in one corner, to visualize the entire changeover. This helps immensely when you need to scrutinize each activity for reduction possibilities. It enables you to examine those actions that have always taken place while the machine was off. The main objective is to convert whatever possible to external time so that the step can be completed while the equipment is still running. Furthermore, the videotape can be taken out of the shop and reviewed in a classroom setting.

It is external (preparation) time that presents the single greatest time-saving opportunities. It would not be overly ambitious to anticipate a saving of 50 percent in total setup time by using heretofore unused time. This entails listing and enumerating all the routine chores that must be done in preparation for the upcoming job.

Up to this point the only cost involved has been that of videotaping. Separating external functions from internal ones is merely a matter of organization, which carries no cost. Even if the internal changes do involve some expense, it will be dwarfed by the benefits obtained. For example, once a stop has been fabricated, which eliminates a changeover step, this time-saving feature is there to use each time the particular die or machine is set up. The cost has been incurred one time.

The saving of time realized through the analysis of internal time, when a particular piece of equipment is running, is constrained only by one's ingenuity. The greater portion of internal setup time is taken up by adjustments and clamping. They become the candidates for study.

It may take only a few seconds to get a die into the bed of a press, but seemingly forever to get it positioned correctly. Most machinery is designed to be infinitely adjustable over a given range. Once the precise setting is located, it is documented. In order to make it possible to locate this setting repeatedly, enhancements must be added wherever required: building guides, cutting notches or grooves, locating stops, or doing whatever is necessary to ensure that the first movement in setting a die is the only one. Each action taken is fully documented into a set of instructions that is kept at the machine. The goal is to duplicate this process again and again, but using increasingly less time.

An analysis of the clamping operation is equally revealing. If one can get an idea of how much force is being exerted and in what direction, the clamping devices can be designed accordingly. Possibly 4 to 6 bolts will do the tie-down job nicely, instead of the 12–16 now being used—and yet another waste is drastically reduced. Consider the type of clamp or fastener. Threaded fasteners are the most inefficient ever devised. Threads strip, nuts freeze, bolts break. And another area reveals itself for scrutiny.

Every activity in equipment changeover or assembly setup is replete with opportunities for you to prove your inventiveness. Your goal should be a 75 percent reduction in setup time, which is fully within your grasp. You must focus on delay time: the elapsed time between the last piece run on the current setup to the first good piece from the changeover. Figure 9–1 illustrates the changeover process.

Another benefit accrues automatically without being planned: an improved learning curve. Under a JIT shop operation enough parts will be run for several days, rather than for the several months required with batch manufacturing. This means that the setup repeats every couple of days instead of every few months. By repeatedly setting up the same process over brief periods of time, the operator becomes adept more quickly.

It must be fully understood that a reduction in setup time is not an end in itself. It is a means to the end of increased throughput. Furthermore, the *total* hours expended in setup activities is not likely to change. There are just going to be a whole lot more setups. One way to envision it is to compare the

**FIGURE 9–1**
**Setup Time Reduction Recap**

| *Steps Taken* | *Amount of Reduction* | *How Accomplished* |
|---|---|---|
| First reduction activity | Reduce by 50% | 1. Separate external time from total setup time<br>2. Eliminate waste in external time |
| Second reduction activity | Reduce by another 50% | 1. Change internal activities to external time<br>2. Eliminate adjustments<br>3. Eliminate clamping |
| Continuing reduction activities | Reduce by another 40% to make total reduction 90% of original | 1. Change the process<br>2. Gradual improvement |

time involved in setting up and running an order for 100 pieces with what it would take to run 10 orders of 10 each given the nine additional set ups.

The original intent of just-in-time manufacturing remains quicker throughput of ever decreasing lot sizes. When you stop and think about it, parts being processed have only two possible end options: Either they are further processed by the next operation, or they go into work-in-process to wait. The goal of inventory reduction can only be served by strictly adhering to the first option. Smaller lot sizes and inventory reduction go hand in hand.

The JIT philosophy actively fosters productivity improvements. You'll recall that we said earlier that there are two components of each JIT activity: predictability and visibility. Consider what impact these have on the renewed confidence that can now be placed in the planning and scheduling areas. You now know what you are going to do and how to accomplish it. You now know when the activity will begin and when it will end. And this puts you back in control.

Once the reductions in setup times have produced more flexible operations, the main building block has been set in place. The next step is to determine what impact this has on capacity and how to use it to achieve a better degree of control.

## PRODUCT LOAD PROFILE

In our discussion of rough-cut capacity planning the product load profile was described as necessary for testing bottleneck work stations in an attempt to verify the doability of the master production schedule.

When first examined, these product load profiles, once calculated—unless there were wide swings in order quantities—were not subject to very much change over time. Their number was suggested to be included as part of the item master file as defined under MRP.

Now, however, because setup times are the major variable in the load profile calculation, reducing these times may directly affect the numbers generated, so the product load profile takes on an added responsibility. Due to the need to load the various production processes and the desire to decrease run quantities brought about by reduced setup times, the delicate balance that exists between plant load and plant capacity could well be in jeopardy. This may tend to distort or even destroy the predictability/visibility features in your system that are being so carefully nurtured.

Several sample calculations can better illustrate this. Returning to the original computation of the product load profile:

$$\begin{aligned}\text{Order quantity} &= \text{1,000 pieces}\\ \text{Average output} &= \text{133 pieces/hour}\\ \text{Setup time} &= \text{3.5 hours}\end{aligned}$$

$$\begin{aligned}\text{Fabrication time} = 1{,}000 \div 133 &= \text{7.5 hours}\\ \text{Setup time} &= \underline{\text{3.5}}\text{ hours}\\ \text{Total elapsed time} &= \text{11.0 hours}\end{aligned}$$

Converting 11.0 hours elapsed time to minutes:

$$\begin{aligned}11.0 \times 60 &= 660\\ \text{1,000 pieces} &= \text{660 minutes}\\ \text{1 piece} &= \text{0.66 minute, or 40 seconds}\\ \text{Cycle time} &= \text{40 seconds}\end{aligned}$$

As the parameters change—that is, the variables of run quantity and setup time resulting from JIT implementation—the balance can be upset. Unless the reductions in setup time and size of the run occur at the same time and in the same ratio, your project could be frustrated for a time. The next two calculations exhibit what can happen when the reduction in run quantity does not keep up with the decrease in changeover time and when it outstrips the setup reduction.

$$\begin{aligned}\text{Order quantity} &= \text{400 pieces}\\ \text{Average output} &= \text{133 pieces/hour}\\ \text{Setup time} &= \underline{\text{2.5}}\text{ hours}\end{aligned}$$

$$\begin{aligned}\text{Fabrication time} = 400 + 133 &= \text{3.0 hours}\\ \text{Setup time} &= \underline{\text{2.5}}\text{ hours}\\ \text{Total elapsed time} &= \text{5.5 hours}\end{aligned}$$

Converting 5.5 hours elapsed time to minutes:

$$\begin{aligned}5.5 \times 60 &= 330\\ \text{400 pieces} &= \text{330 minutes}\\ \text{1 piece} &= \text{0.83 minutes, or 50 seconds}\\ \text{Cycle time} &= \text{50 seconds}\end{aligned}$$

Here the cycle time drops to a low figure, but fabricating time is still quite high. The key figure in the calculation, total elapsed time, shows that benefits are not being realized and results are likely to be discouraging. The solution is to reduce run quantities until cycle time equals 40 seconds.

The other condition shows the exact opposite effect, yet the results are nearly identical:

Order quantity = 800 pieces
Average output = 133 pieces/hour
Setup time = 1.0 hour

Fabrication time = 800 ÷ 133 = 6.0 hours
Setup time = 1.0 hour
Total elapsed time = 7.0 hours

Converting 7.0 hours elapsed time to minutes:

$7.0 \times 60 = 420$ minutes
800 pieces = 420 minutes
1 pc. = 0.53 minute, or 31 seconds
Cycle time = 31 seconds

When you drop the run quantity without a commensurate reduction in the setup time, the result is needless use of available capacity. In other words, it is probably cheaper and more productive to leave the run quantity at the original 1,000. With a cycle time of 50 seconds, it is 10 seconds longer (4,000 seconds longer for a run of 400 pieces). This uses over an hour (4,000 ÷ 3,600 = 1.11) more time than does the original quantity and should be avoided. One of the common pitfalls in Just-In-Time implementation is impatience—trying to take too big a bite at one time. Solution: more concentrated effort on setup-time reduction before cutting run quantities.

Take a look at the final calculation, where the setup reduction is coming along well and the drop-in run quantity is keeping pace.

Order quantity = 400 pieces
Average output = 133 pieces/hour
Setup time = 1.0 hour

Fabrication time = 400 ÷ 133 = 3.0 hours
Setup time = 1.0 hour
Total elapsed time = 4.0 hours

Convert 4.0 hours elapsed time to minutes:

$$4.0 \times 60 = 240 \text{ minutes}$$

$$400 \text{ pieces} = 240 \text{ minutes}$$

$$1 \text{ piece} = 0.60 \text{ minutes, or } 36 \text{ seconds}$$

$$\text{Cycle time} = 36 \text{ seconds}$$

The cycle time of 36 seconds is almost equal to the original cycle time of 40 seconds. This means that excess capacity will not be required to accomplish this task, as in the previous example; nor is it likely to add unduly to the work-in-process, as would probably be the case with the first example. Both setup time and run quantity have been reduced by like proportions, while effecting a 60 percent reduction in total elapsed time. Furthermore, this has been realized without causing any adverse reactions elsewhere in the system. In other words, to achieve this benefit does not require a trade-off someplace else.

A final point is well worth noting. Throughout this activity the average output has remained at 133 pieces per hour. This illustrates a point central to JIT philosophy. The firm's resources remain constant. What has changed is the way in which these resources are used. The way to add value is not to change resources, but to add value, eliminate the causes of waste that have crept into the system and accumulated over time.

## GROUP TECHNOLOGY

There are two concepts involved in group technology having to do with combining production characteristics so that manufacturing or assembly functions are more directly channeled into accomplishing an assigned task. By far the most prevalent is *cellular manufacturing,* often called *U-Line Layout.* Actually the "**U**" encompasses more than the simple idea of putting the starting and ending positions adjacent to one another to take up less floor space. It is meant to convey a significant alternative to the way things have always been done. The logic behind it is to create a more or less natural grouping of things, either products or processes, that are similar and that lend themselves to a cellular setting where the operations of value-adding can be accomplished in a single continuous processing function.

The operations could combine several processing steps into one major step or into the completed product. The cellular configuration certainly suggests a **U** shape, but it could just as efficiently be a right angle or even a straight line, depending upon the needs of the individual user. Figure 9–2 points out such a relationship.

This figure illustrates the working relationship that might exist within and between work centers in, say, a small metal fabrication shop.

In a typical facility that carries on conventional batch processing, the shop-equipment or work centers are intended to carry on an individual task at a given time. After completing this task, the operator will set up and run another work order, the material for which is queued up behind, waiting its turn. All work centers in the shop operate in the same fashion.

Examination may reveal that products that make up a given class or family of items routinely follow the same circuitous route through the processing steps each time a new manufacturing run comes due. Each work center is completing a different processing step, but maybe at the same time, with the various operations oblivious to the activities of one another.

With cellular manufacturing, it becomes more logical to group these processing centers along the same work path uti-

**FIGURE 9–2**
**Batch-Processing Layout**

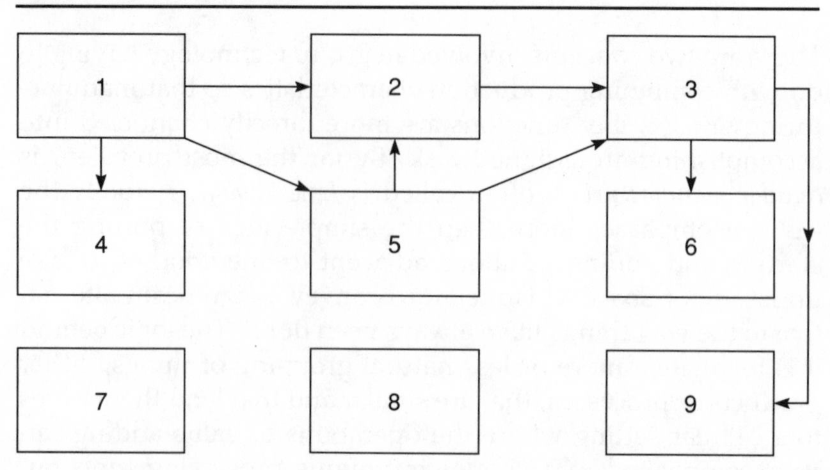

lizing the group technology (U-line) concept. In this way the products take on a flow characteristic, as opposed to the starting, stopping, queueing, and restarting of traditional modes of manufacturing.

The assigned tasks within and along the cell will not change to any great extent, but the idea of teamwork has now been introduced, so the objective is to balance each step of the operation so that a flow can be maintained. The result is a balanced rate, in sync with either the succeeding operation or the customers' due dates. Figure 9–3 distributes these nine work centers around a U-line layout.

The human element in the cellular concept is perhaps its most dominant feature. Throughout the entire spectrum of JIT strategies, people are of paramount importance. Group technology is no different. The division of assigned tasks can only be undertaken by those involved in the process. Their interaction makes it happen.

In addition, speedup or slowdown in shop activity is usually not lightly proposed. Under the U-line layout any increase, whether a momentary thing or a significant pickup in business, can be handled by adding one person to the crew and realign-

**FIGURE 9–3**
**U-Line Layout**

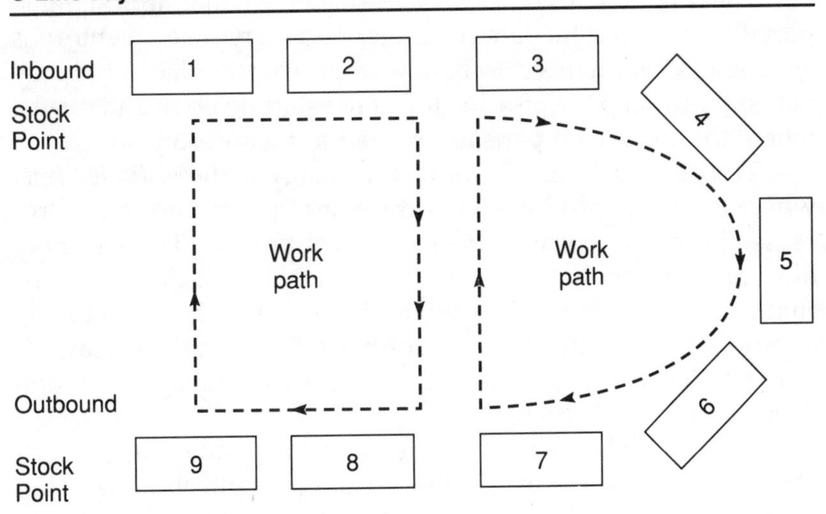

ing responsibilities. The reverse is handled similarly: When a drop in demand or a slowdown is called for, a person can be dropped from the line and assimilated somewhere else. The balance remains unaffected, but the flow continues at a faster or slower pace.

Operating a cell uncovers two unexpected benefits. First, it provides an ideal place for cross-training workers. Typically jobs are rotated regularly which serves to dampen the tedium of humdrum jobs. Second, the learning curve improves, so that as the group operates more as a team, they become increasingly efficient.

Many of the former causes and costs of waste have now been eliminated. The queue of parts normally waiting behind each work center is gone. The work center is now processing the precise amount required to balance the flow. This is one of the more obvious advantages, and it happens as a result of another, completely separate activity. The cellular line configuration simply provides no space for the storage of in-process inventory.

Another benefit of cellular manufacturing is improvement in quality. As parts flow along the work path, each one is handled individually, probably more than one time. Feedback from a deficiency in the product or process is instantaneous, allowing for a correction to be made at once. This precludes the line from producing an entire lot of unacceptable parts.

The final benefit comes from the material handling and transportation activities that are no longer necessary. The inventory is gone and so is the need to haul it from place to place. It is only necessary to keep the pipeline full at the starting point and maybe where the completed parts are moved to their destination.

The second form of group technology is the *dedicated line,* where the same products with the same specifications are processed in the same way in the same work centers. The assembly line, or work center configuration, once established, remains that way permanently. This would be used by a firm producing a proprietary product that accounts for the bulk of (say, 70 percent) its sales volume. Obviously this is a less common form of group technology.

The two aspects that permeate all JIT strategies, predictability and visibility, are evident in the group technology concept. The regulated flow allows start, stop, and changeover to be scheduled with precision, while visibility within the cell is total and immediate. Because the plant layout has been realigned and work-in-process has disappeared, visibility over the entire process is greatly enhanced.

## SUPPLYING THE INBOUND STOCKPOINTS

If the starting operation begins with one piece that is added to as it flows toward completion, the supply problem is a simple one. Either the entire order quantity can be accumulated adjacent to the beginning operation or, if this is not practical, a certain quantity can be replenished at specific intervals.

On the other hand, if there are a combination of components that will be welded, bolted, riveted, or otherwise fastened together, the problem becomes one of visual control. An area that is easily accessible to both supplying and using departments must be designated as Component Storage. Within this area lanes can be designated as the replenishment spot for a specific fabricated part. The lanes could be painted on the floor the width of a skid or even of the part itself, if it is too large or unwieldy to fit into a container. These lanes could be marked with a simple and understandable (say, alphanumeric) designation.

The entire area would be the replenishment point for all the manufactured components, with the quantity of each easily visible. The replenishment method is a simple two-bin system. Some manner of signal marker would be used to indicate when the stock is drawn down somewhere near the lead time required to manufacture more. The signal marker would then be affixed to a marker tree, signal pole, or some similar structure with high visibility to indicate that a replenishment quantity is needed. The next two figures, 9–4 and 9–5, show how these might look.

As setup times continue to decrease and smaller replenishment lots are produced, the supplying and using persons, among themselves, can decide on a replenishment scheme. For example, if a part is fairly common and frequently used, a two- or three-day quantity could be the standard reorder. Because the system of replenishment becomes strictly tied to actual usage, it qualifies as a pull system.

**FIGURE 9–4**
**Supplying Inbound Stockpoints**

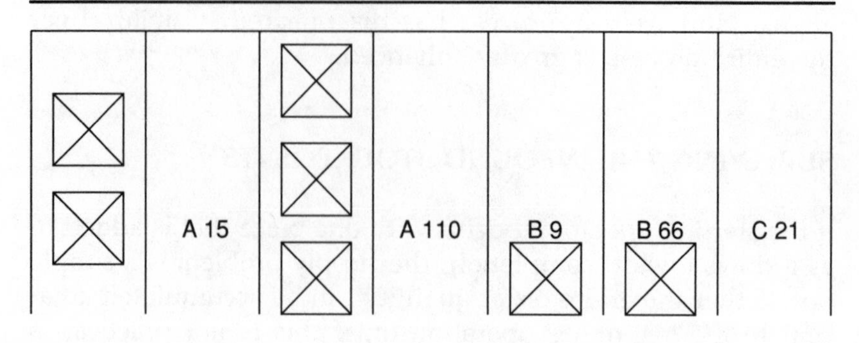


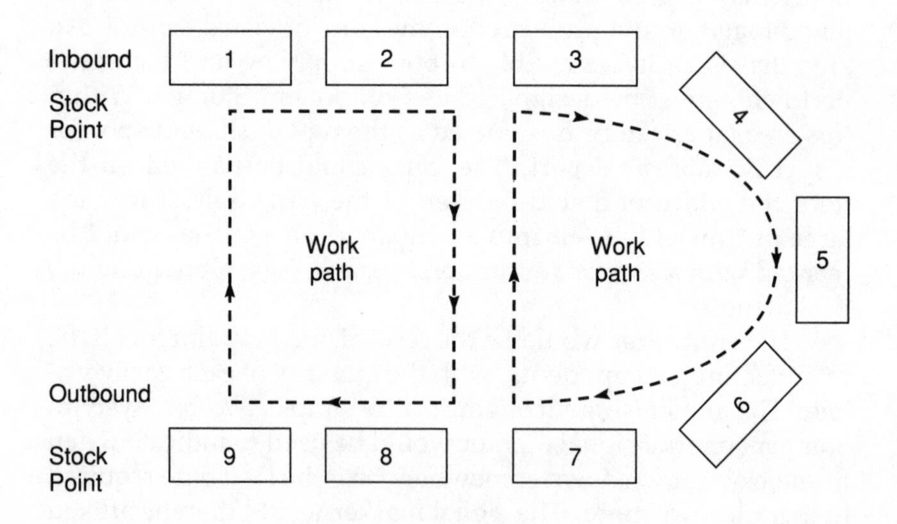


## USING THE SYSTEM

Up to this point most of your effort has been in the direction of leveling the load. Obviously this cannot be done until the run sizes are smaller and tied to what is needed and can be consumed rather than to some arbitrary quantity that tends to increase inventories.

**FIGURE 9–5**
**Outbound Stockpoint with Warning Signal Marker***

Diagram of outbound stockpoint with warning signal marker

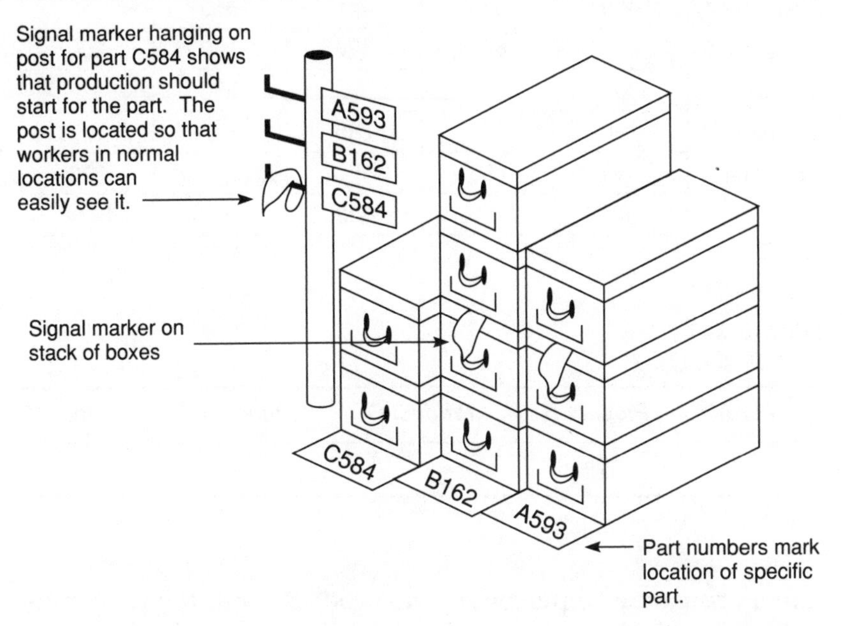
Signal marker hanging on post for part C584 shows that production should start for the part. The post is located so that workers in normal locations can easily see it.
A593
B162
C584
Signal marker on stack of boxes
C584
B162
A593
Part numbers mark location of specific part.

Note: These are more than reorder point systems. The intent is to progress beyond the use of the signals, not use them indefinitely.

Source: Robert W. Hall, *Zero Inventories* (Homewood, Ill.: Dow Jones-Irwin, 1983), p. 51

The undisputed primary document that everything builds upon is the master production schedule (MPS). This is a detailed account of what all of your customers want and by when, so it is the organization's responsibility to make it happen. This is accomplished by the final assembly schedule (FAS).

The FAS is the current portion of the MPS that details the quantity and the timing of each week's assembly of end items. In order to accommodate customer requirements, a leveled FAS is mandatory. This is especially true when customer demands are random occurrences. Figure 9–6 depicts a breakdown that a typical computer printout might show by product type and by customer on an annual basis. Using a Pareto analysis, five cus-

**FIGURE 9–6**
**Customers' Annual Usage**

| | Product A | Product B | Product C | Product D | Product E |
|---|---|---|---|---|---|
| Customer 1 | 450 | 350 | 40 | 650 | 125 |
| Customer 2 | | 600 | 430 | 160 | 280 |
| Customer 3 | 800 | | 82 | 550 | |
| Customer 4 | 120 | 116 | 280 | | 1,163 |
| Customer 5 | 450 | 650 | | 200 | 200 |
| Totals | 1,820 | 1,716 | 832 | 1,560 | 1,768 |

**FIGURE 9–7**
**JIT Leveled Load**

| *Product A* | *Product B* | *Product C* | *Product D* | *Product E* |
|---|---|---|---|---|
| 35 | 33 | 16 | 30 | 34 |

tomers consume 80 percent of the product, with the remaining 20 percent going to a variety of smaller users. It should be noted that products must be listed in units because dollar volumes would be meaningless.

The next step is to break each product down into 52 periods of equal size, as shown in Figure 9–7. The FAS for each week would appear as shown.

These 148 units plus another 30 (the remaining 20 percent) become the target for you to aim at. This forms the basis for the weekly assembly schedule. The ultimate objective, of course, is a *daily* schedule, and even this can be improved upon. But getting there can only be accomplished through a long succession of gradual improvements.

It is certainly apparent how this schedule differs from the more traditional form of assembly scheduling. It had always been thought that the key to productivity lay in the cost-effective approach of putting together a rather large number of identical parts. The storing of these parts would cover current demand while another large group of items followed in sequence. This would continue through the entire product structure, and then the cycle would be repeated. The typical schedule would look similar to Figure 9–8.

It would not be at all abnormal to become truly efficient and run up several hundred of one or more of the items when their assembly time came due. However, scheduling in the manner shown in Figure 9–8 was the only way to make it through each of the product groups every month. And it did not do that bad a job of serving the customer, although it would be poor by today's standards. It is probably a safe bet to assume that this volume and mix of production was accompanied by many overtime hours, precipitated, for the most part, by countless schedule changes.

This is still another instance where operating under a Just-In-Time scenario is advantageous. For example, you cannot run any more than 35 units of Product A each week unless they are scheduled that way in advance. The parts for the overrun are simply not there, and another lot will not be available until next week. There is no work-in-process to help draw from because this has become extinct. Furthermore, you are confident you will have the components for 35 units of Product A, not only this week, but every week; and not only Product A but Products B through E, because the program includes working toward a target of zero defects.

In any event, there may be occasions throughout the year when a firm may experience slack periods or very busy ones. Management will be forced to choose whether to build up some

**FIGURE 9–8**
**Monthly Schedule for Conventional Manufacturing**

| | Product A | Product B | Product C | Product D | Product E |
|---|---|---|---|---|---|
| Week 1 | 152 | 8 | | | |
| Week 2 | | 135 | 25 | | |
| Week 3 | | | 44 | 116 | |
| Week 4 | | | | 14 | 146 |

inventory in advance of the peak months or increase capacity during these times. These are management decisions, not Just-In-Time strategy decisions. A leveled load is not without its own set of considerations. Greatly reduced lead times will go a long way toward softening the effects of the sharp peaks and valleys that so often characterize demand. But there are still enough times when you have to cope with a sudden influx of 10 pounds of demand while holding only a 5-pound bag.

A leveled load in the FAS must be preceded by a level load in the feeder processes that lead into it. This can only be accomplished through smaller lot sizes of greater product variety. Because of capacity constraints, this cannot happen without equally sizable reductions in setup times. This is a rule that can never be forgotten. Smaller lot sizes are a function of setup reduction and will never come about in its absence.

A level schedule in final assembly is dependent on better throughput in upstream operations. Better throughput comes from the proper mix of smaller lots, which in turn are made possible because of reduced setup times. So the correct way to proceed is to reduce standard lots in the same percentage as setup time is reduced and test it. If it works, you have made a significant breakthrough. If it does not, do not get discouraged and give up. Review your steps and retest. Eventually it will succeed, and you will have gained some important insights in the process. These will be a valuable guide to your continued efforts as you forge ahead.

## THE PULL SYSTEM

The epitome of sophistication in a Just-In-Time system is the pull mechanism. That it took Toyota 25 years to evolve their Kanban system is a testimony to this. The pull mechanism is the final implementation activity. Everything must be in place and operating satisfactorily in order for it to achieve its specific result.

There are three points to bear in mind regarding the pull system. Practically the first thing anyone ever brings up in a discussion about Just-In-Time is the elements of a pull system, but this can be dangerous because the pull system is emphatically not one of the first JIT strategies one employs. To have that clouding one's thinking or, even worse, setting a timetable for its implementation may be an invitation to disaster.

The second point is a corollary to the first and is equally clear. The pull mechanism is the last element to be put in place. It is the culmination of all those activities that have preceded it, and it must be a natural evolutionary step that is taken because everything indicates you are ready for it.

Finally, although it is the last implementational step, the pull mechanism by no means suggests that gradual improvements have run their course. For just as the objective of the organization is to operate in perpetuity, the same timetable applies for gradual improvements.

It was noted earlier that conventional manufacturing is typified by a push system—that is, each lot of parts is pushed around the shop to various work stations as the routing sheet indicates. Under such an environment, the impetus to perform an activity of either production or assembly is derived from a requirement in the form of a shop order. The underlying principle guiding this is that other similar requirements for the same end product will be undergoing their own processing in congruent fashion and that all will come together, more or less simultaneously, to satisfy a current need. In other words, to be an activity, there must be a requirement and this requirement will carry a completion date. The summation of the requirements along with their proposed completion dates comprise the production schedule.

The push system, then, is a scheduling mechanism based on dates. The pull system, on the other hand, is a signaling mechanism based on timing. It is the signal that activates something happening in either production or assembly. The signal means to supply a need. The pull system, therefore, is a signal going from the point of use back to the point of supply (production or assembly), signifying that it is time to fill a need.

Therefore, the critical difference between the two systems is that the pull system fills a need and the push system fills other quantities. The other difference between the two systems is one of movement. A push system is punctuated by starts and stops interspersed with long setup periods (or maybe something worse), all geared to some future date of need. A pull system employs basically the same functions and uses the same tactics, but the periods of interruption are quite short, allowing for more product variety of a balanced nature, where the processes tend to take on more of a flow character and supply an immediate need. The common sense rationale underlying this basic concept is conveyed by Figure 9–9.

It should be obvious from this description that involving one's self in a pull system requires intricate preparation. This is why it is always done last. Because it is based on timing, it embodies two features without which it would be impossible to operate: uniformity and synchronization. The vibrant manner that typifies the environment within which such a system operates fairly demands their existence. A number of using points continually signaling a larger number of supplying points, each with its own specific need to be filled, suggests a precision tantamount to that of a drill team.

Also suggested from the operation of a pull mechanism are four functions that must be in place if the system is going to work. The first is quality, which permeates the JIT strategy throughout. In addition to being uniform, this quality must also provide superior value. It must be a continuous, conscious effort everywhere in the entire system.

The second function is keeping a simple flow pattern where product movement and wasted motion are held to a minimum.

**FIGURE 9–9**
**Fixed-Volume Pull System**

Concept of fixed-volume pull system

If the feeding process has nowhere to put inventory except in the pipeline, it can only produce and fill the pipeline when an empty space appears.

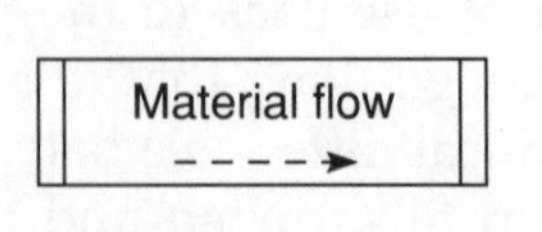


Removal of material for use causes an "empty space" in the pipeline. Material moves down the pipeline to leave an empty space at the feeding end.

Source: Robert W. Hall, *Zero Inventories* (Homewood, Ill.: Dow Jones-Irwin, 1983), p. 40.

Strongly indicated by a compact plant layout is the third feature: communication and feedback. In addition to being immediate and instantaneous, this flow of information must also be accurate. And accuracy is greatly enhanced by the proximity of the using and supplying processes. Fourth are the practices of housekeeping and preventive maintenance, which are taken up next.

Once thrown into forward gear, the pull system is paced by its two hallmarks, uniformity and synchronization. This points out yet another good reason for a high state of preparedness: Once set in motion, the system cannot be stopped. Its beauty lies in the ability to perpetuate itself.

## HOUSEKEEPING

Cleanliness for the sake of cleanliness is an admirable idea, but cleanliness to reach a stated objective is entirely different. The former is a favorable impression that management wishes to convey of its overall attitude toward the work force and their workplace, while the latter imparts more than just attitude.

Housekeeping within the Just-In-Time strategy takes on added significance. The work area must be free of obstructions as well as clutter. One of the prime objectives of restructuring a plant layout is to improve visibility, which is essential to the communication and feedback routines.

The same thing is true of the material flow process. Obstructions are impediments to the value-adding process and must be removed. Maintaining a steady pace throughout the entire shop is so much easier when work stations are plainly visible to each other.

## PREVENTIVE MAINTENANCE

Earlier we stated that housekeeping and preventive maintenance, as JIT tactics, were in no way new, but that housekeeping has always been a common activity because it is so highly visible, while preventive maintenance, by and large, was only paid lip service. And why not? It stood in the way of "productivity." That it was performed at all was due more to management's insistence than any perceived need.

If one wishes to operate within a pull system, preventive maintenance must rank on a par with reduced changeover time, small lot sizes, and every other technique that is considered indispensable to a JIT operation. The rules are no different than they were some 30 years ago: A prescribed time is set aside before production begins for each operation, each shift to perform a checklist of operating conditions. Failure to perform these checks is tantamount to lying and should be subject to the most severe disciplinary measures. The other part of preventive maintenance is scheduled shutdowns. Each piece of production equipment must receive a regularly scheduled overhaul. Whether it's a major or a minor overhaul depends upon your particular set of circumstances, but it must be removed from service so that the work can be performed.

When these ideas and techniques are in place and working, think of the built-in predictability your system will then possess.

## ELEMENTS OF PEOPLE MANAGEMENT

Under a JIT scenario the relationship between management and its employees undergoes a unique transformation. One of the most remarkable manifestations of this change is that as gradual improvement occurs, the employees themselves provide the impetus for change. Everyone is well aware of the direction of the company, so there is no need to seek approval for initiating a change that is seen as sensible. Supervisors, thus, can spend less time supervising and more time acting as facilitators.

Obviously this is an abrupt reversal of the traditional form of employer/employee relationship. The old "you ain't gettin' paid to think" style of management must be silenced forever. The statement has been made that the Japanese hire people from the neck up, while the United States hires people from the neck down. This statement may or may not be accurate, but it serves as a warning. If trust and cooperation—so vital to a highly motivated workforce—are not engendered and reinforced by managers, they will find themselves missing one of the basic resources necessary to eventual success. One important goal is to shed the "one-who" and its stepchild the "c.y.a. syndrome" and the negative images they help create. Faultfinding must be replaced with fact-finding.

## TEAMWORK

What are recognized and rewarded are improvements in customer service. These can be achieved only through team effort. The individual strives for excellence, but must be cognizant of membership in the team. This tends to promote group loyalty and response to peer pressure. Each person contributes to the team effort, and each is rewarded as a valuable member.

## SKILL DIVERSIFICATION

The concept of teamwork is a great balancer within an organization. Another balancer is training each employee to perform multiple tasks—what is sometimes referred to as a *multifunctional workforce.* This is more easily achieved in a repetitive environment, however, which is consistent with a group technology approach. The idea is to rotate jobs regularly among individuals in their immediate environment. Then the environment can be expanded to encompass other responsibilities. This provides two benefits to the company. First, the more people there are qualified to perform a given task, the less impact there will be when an employee is absent or decides to leave. Second, skill diversification greatly increases the likelihood of suggestions for improvement. It is usually of more immediate benefit to the employee, though, because many firms pay according to how many jobs each employee can do.

Unfortunately, this diversification conflicts with the traditional position of trade unions, in which a certain pay scale for a particular job classification was almost a sacred trust. Recently, however, this position appears to have softened considerably from their more militant days as they face global competition for jobs.

## SCANLON PLAN

Ever since the original caveman entreprenuer built the first Stone Age factory to mass produce wheels, managers have continually searched for that one particular thing that motivates workers to operate at their productive peak. This generally has taken the form of money, and no matter what it is called, it is a bonus. There are literally hundreds, maybe even thousands, of various bonus plans in existence—all the way from the simple Christmas or year-end bonuses to systems so intricate that most employees don't understand them.

The track record of bonuses is mixed. Many, like the once-a-year bonuses, are not tied to anything specific and often come to be expected as a matter of course. Once begun, they are difficult to discontinue unless some other plan takes their place. In any event, as a motivator they are relatively useless and their impact on productivity is minimal. Direct labor productivity bonuses only serve to encourage production whether it is needed or not. It is a self-defeating plan because most are based on standard hours and have no way of deducting for parts needing rework or scrapped parts. Many bonus plans, maybe even most of them, have drifted far afield from their original intent. They pay the employee for individual productivity or efficiency based somehow on profitability, which will motivate them to be even more productive, earning still bigger bonuses, etc. Individual bonuses are diametrically opposed to the JIT crusade. They do absolutely nothing to improve service to the customer; they have no real basis in promoting quality; and to some extent they are, in themselves, wasteful.

Productivity sharing plans are now being called *gainsharing plans,* and the most highly regarded of these is the Scanlon Plan which has been around since the 1930s. The basic premise proposed by its architect, Joseph Scanlon, is that the employee on the job is the most familiar with that job and best knows how to avoid waste, improve efficiency, reduce costs, and improve the quality of the product. The challenge of the Scanlon Plan is to get companies to open themselves up to ask for help and, equally, to get the employee to be willing to share his or her capability and knowledge with management.

The philosophy underlying the Scanlon Plan is that the organization should function as a single unit, that workers are capable of and willing to contribute ideas and suggestions, and that resulting improvements should be shared. The participative style of management of the Scanlon Plan benefits both management and employees. The motivating force, as with all plans, is a bonus, but it is tied to overperformance to goal or a set of goals whose parameters are arrived at jointly by the company and the workers. The employees further benefit because each feels that he or she is an important part of the corporation. They know they can contribute valuable insights to a more effective work environment and that when they are asked, they will be listened to.

The company gains in three ways. First, overperformance to a goal is distinctly more profitable. Second, it is truly productivity-based, the cost avoidance or cost saving being quantifiable. And third, high job satisfaction and motivation are directly attributable to low absenteeism and turnover.

The elaborate structure of the Scanlon Plan is one more feature of its uniqueness. It is made up of two-tiered production and screening committees. Production committees are department- and shift-based and also include both clerical and office employees. Production committees normally consist of two to five hourly employees elected by their peers plus a supervisor or manager. Their responsibilities are specific: to encourage idea development and evaluate employee suggestions. Supervisors can veto any suggestion, but the employee can appeal to the higher-level screening committee.

The screening committee is likewise made up of hourly and salaried employees. In some shops, the workers' representative would typically be the steward or some other union officer. The responsibilities of this committee are more general. In addition to adjudicating appeals, they also rule on suggestions that cross jurisdictional boundaries or exceed the cost guidelines of the production committee. All concerns of a general nature are considered at this level, not the least of which is a review of the monthly bonus calculations. The internal communication provided for in the plan permits the free flow of suggestions, as shown Figure 9–10.

The Scanlon Plan in its purpose and implementation closely parallels quality circles, but there are significant differences. Quality circles do not generally include bonuses as a motivational force. Further, the employees are rather restricted to their immediate work areas, and the committees aren't as permanent. Both systems, however, share the participative style and humanistic orientation so necessary for the success of each.

Too many managements shy away from the Scanlon Plan because they find it difficult to involve subordinates in operating decisions. For example, many will hesitate to share profit-and-loss or other financial information of the company. For these companies it is good that they stay away from the Scanlon Plan, because sharing and trust are absolute necessities for its success.

**FIGURE 9–10**
**Scanlon Plan Suggestion Flow**

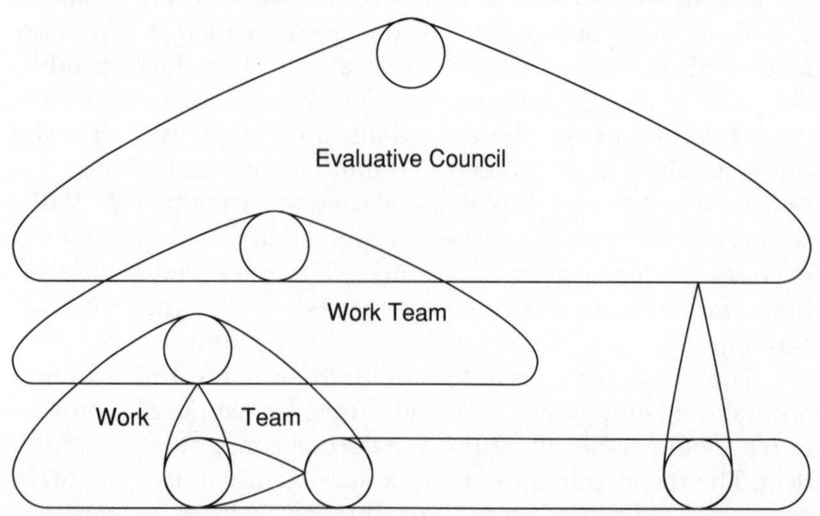


Source: Hy Bomberg, "Productivity by the People," *Management World*, February/March 1983, pp. 22–24.

## SUPPLIER NETWORK

There are three conventional purchasing maxims:

1. Never tell the supplier anything more than he or she absolutely, positively has to know.
2. From the supplier's viewpoint: Never tell the customer anything more than he or she absolutely, positively has to know.
3. Always buy from the vendor who offers the cheapest price irrespective of other attributes.

Although these are written with tongue in cheek, they are no joking matter. How many times have you seen them followed almost to the letter? As with every other segment of Just-In-Time, a new mental attitude must be evident here as well. Vendors are not the enemy. They are not scamming confidence men out to bilk the customer nor to blackmail him with "leaked" information—but is that not precisely how he is being treated?

## A MODERN APPROACH TO VENDOR REACTIONS

That these conventional wisdoms do not make any sense and actually run counter to one's own best interests has never been seriously considered. They are condoned by the oldest reason in the world: "But we've always done it that way."

Suppliers should be considered partners. The ties that bind manufacturer and customer should be precisely the same as those between manufacturer and supplier. After all, their goals are identical: to share in the pot of gold at the end of the rainbow—but not at the expense of one another; with the connivance, if you will, of each other so that the reaction of one complements the actions of the other.

There is no percentage in acting as though a state of war exists between you and your supplier. For just as your own employees have come to feel that through their own expertise they are helping to fulfill the goals of the company, so should the vendor feel that, being an expert about the product supplied, he is playing a similar role in reaching those goals. The supplier has an economic interest in the success of his customer.

Once you have come to realize the changed relationship with your suppliers, you have to then decide which vendors they will be and whether you prefer single or multiple sourcing. These are separate problems because of circumstances, and so they will be dealt with separately.

It is not within the scope of this book to set forth the complicated rules of vendor selection, which is the topic of many purchasing articles and texts.

## A NEW LOOK AT SOME OLD CONCEPTS

But several things stand out that bear mentioning. The traditional triumvirate that are included in all value analyses are, in order of importance, price, delivery, and quality.

Delivery is one item in the supplier's history that is simply a matter of record. And good or bad, that record is not likely to change.

The same cannot be said of quality and price. Quality is the single most critical determining factor in keeping a vendor. Unless a vendor can prove that it is capable of consistently (which means very nearly always) supplying you with nothing less than your minimal standards, it doesn't matter what else it does for you. You cannot retain that vendor, even if it's your brother-in-law.

Price is the most discussed yet least understood of the three factors. It is the easiest one to quantify (a vendor's price is lower or it's not), but the price of a unit is not the only financial consideration. Value analysis, employing rating factors, can provide pretty good indications of a vendor's relative worth. That is why it may be shortsighted to focus on selling price when a more telling number would be the total cost of procurement.

Furthermore, beyond the current selling price, what efficiencies and cost improvements is this vendor likely to effect in order to reduce cost? Are there future cost reduction projects, and what can you, the customer, do to help? Again the focus should not be on price, but on costs.

Moreover, the terms themselves, price and cost, are not the same. Price is just one element of cost. It is evident only at the time of acquisition, while costs continue throughout. In addition to procurement costs there are carrying costs and life cycle costs, the latter including part failure, other rejects, and rework.

Finally, experience will bear out in many cases that the vendor that scores acceptably in most or all of the selection categories will also be found to have the lowest price. Even in cases where this is not true, negotiations may bring out advantages to the vendor that may lead to lower prices.

A vendor must be looked at in terms that will automatically exclude it. An unacceptable quality rating will put it out of consideration, irrespective of its rating in other categories. If it passes the quality test, it must also pass the delivery test. But a poor rating—or even only a fair one—in either of these categories is a basis for exclusion. Your production lines must not be subject to interruption because of either quality or delivery problems if Just-In-Time is to be a reality. Only after this is done do other vendor performance rating or selection categories, including price, come into play. The quality and delivery categories must retain veto power.

## Single versus Multiple Suppliers

Any discussion of Just-In-Time will include sole-sourcing as one of its basic principles. This is true even though there are plenty of common sense reasons for having two or more vendors. They run all the way from what to do if the vendor burns down to not allowing a single vendor to wield a bigger axe. These are certainly valid considerations but still the arguments in favor of sole sourcing are more compelling and their effects further reaching. For example, alluding to the plant burning down, if two vendors are sharing 50-50 the volume of a given commodity, and one plant burns down, there are no assurances that the other vendor can automatically step in and handle the extra burden, especially in the short run because the need will be the greatest during that time. Uniform quality is an excellent reason for sole-sourcing, especially where the vendor has implemented process control quality. Where several sources have an acceptable level of quality but their products are not identical or even interchangeable, this may cause scrap or control adjustments when switching lots of material during a production run.

Another advantage for using a single vendor is the reduction of freight costs. Having several vendors share a common load will reduce costs even more. One final advantage, now more recognized, is the special terms that two organizations can agree to and that would never be possible with multiple sourcing. For instance, the reduction in paperwork and recordkeeping can yield real dividends. Think of all the paperwork involved in placing and receiving an order. How much of it is really necessary? Purchasing has been using blanket purchase orders and releases for years. This just carries it to the next logical step.

Receiving inspection is another senseless waste, and can be eliminated if the vendor is practicing quality control at the source. Most of the time it is merely checking counts, which often costs more than any missing parts would be worth. Unitized loads, reuseable containers—the list is virtually endless. How much value is added by adhering strictly to control procedures?

## RECEIVING, STOCKING, AND USING IN A J-I-T ENVIRONMENT

The next issue concerns receiving a one-day supply every day. Unless one has infinite receiving capabilities, some common sense rules must come into play. If you are using an assortment of 200 nuts and bolts a day, it is pure fantasy to expect your hardware supplier to show up on your doorstep every day with just the prescribed amount.

Most manufacturers purchase 30–70 percent of their requirements on bills of material. It's likely that at any given time they have scheduled 50–60 individual part lots. Even if they could receive this many loads during a day, total confusion would reign. The logical solution is to break them down into an ABC classification using a Pareto analysis. In this way critical parts could be identified as A parts, with closer controls exercised on them. These parts would thus be on a more frequent replenishment schedule.

At the other end of the spectrum would be C parts, lower-cost items, generally hardware, that are stored in bulk at the using stations, controlled by the two-bin method, and with replenishments going directly to the spot where they are used. These looser controls are quite appropriate. Usage, replenishment, and reorder signals are all controlled by the individuals directly involved, the users. And because they are not subject to very much pilferage, they need not be stored in a stockroom.

The B category is reserved for items that require special consideration, regarding not cost or usage, but rather ordering, receiving, or storing. Examples of this would be packaging supplies with high square-footage needs and certain chemical compounds that require special handling or are subject to deterioration.

The final aspect of getting the vendor involved in the customer/supplier relationship is cooperation. There are only two ways this relationship can be conducted. If the vendor already has a Just-In-Time program under way, cooperation is guaranteed. If it does not, it may feel pressured into an affiliation that is not profitable and its cooperation will be involuntary. It's obvious that there are no easy solutions to this problem. If the supplier is close by, an increase in business may seem to be a beneficial trade-off for more frequent deliveries. But if it is several hundred miles away, it will feel it is being forced to carry your inventory. These feelings are justified because, while backup inventory languishes in a warehouse, no value is being added, but cost is.

It is a common conception that in order to elicit cooperation from a vendor, each must benefit equally from the new relationship. Until each of the critical-component suppliers has established a just-in-time strategy of its own, the benefits will not be shared equally. This is likely to engender a rocky relationship until some compromises can be negotiated.

## SUMMARY

"If you sell it every day, you should make it every day" is a maxim that is sometimes used to describe Just-In-Time. To be sure, this is quite unlike the way you have been used to conducting business up to now. The changes your organization will undergo to accommodate this transformation will be dramatic, to say the least.

It may be that the most difficult nut to crack is employee involvement. That is why there is no way to overstate the need for employee education and training—and not just once, but again and again, in classroom get-togethers and out on the shop floor. Without this employee involvement your program cannot succeed, and you cannot earn this cooperation unless the workers know what is expected of them. With every success you achieve, your goal becomes that much easier to achieve.

One practice you should shy away from is putting time constraints on accomplishing certain goals. The changes themselves are unsettling enough without putting timetables on them that may push them faster than knowledge can be learned and skills acquired. This inevitably leads to finger-pointing and blame-laying, where departments tend to compete with one another rather than complement each other.

In order to get there from here requires a combination of a new way of thinking and charting a course for implementation. It must be fully understood that this course may have to be abandoned and recharted—and maybe more than once—before some positive change is perceived. Throughout this period, however, insights will be gained about your organization and its people that will prove invaluable as you progress through time.

Just-In-Time means not being able to say something will not work until you have tried it—and failed. Again!

# PART 5

# THE AGE OF INFORMATION

# CHAPTER 10

# MANUFACTURING IN THE AGE OF INFORMATION

A number of writers have expressed the notion that the United States and the other highly industrialized countries of the world have entered the Age of Information, which means that the commodity in greatest demand—and therefore, in greatest supply—is information. It is the purpose of this chapter and the next to measure what impact this will have on manufacturing.

## FACTORY OF THE FUTURE

Most experts who have reflected on this subject agree that factories will continue to see automation becoming more and more integrated into the manufacturing process. This will be followed by an attempt to put all the building blocks together to arrive at a total information system called computer-integrated manufacturing (CIM).

The logical culmination of this effort is the so-called *dark factory*, where manufacturing is totally automated and depopulated of human beings. The machinery will be "smart;" that is, if a piece of equipment develops a problem while it is running, it will be capable of diagnosing the problem, shutting down, and repairing itself, all without the need for human intervention. How far into the future before this can be realized is hard to tell, but it was considered science fiction a mere 50 years ago. An artist's concept of such an automated factory is shown in Figure 10–1.

FIGURE 10–1 Factory of the Future as illustrated by Munck Automation Technology, Inc., Newport News, VA.

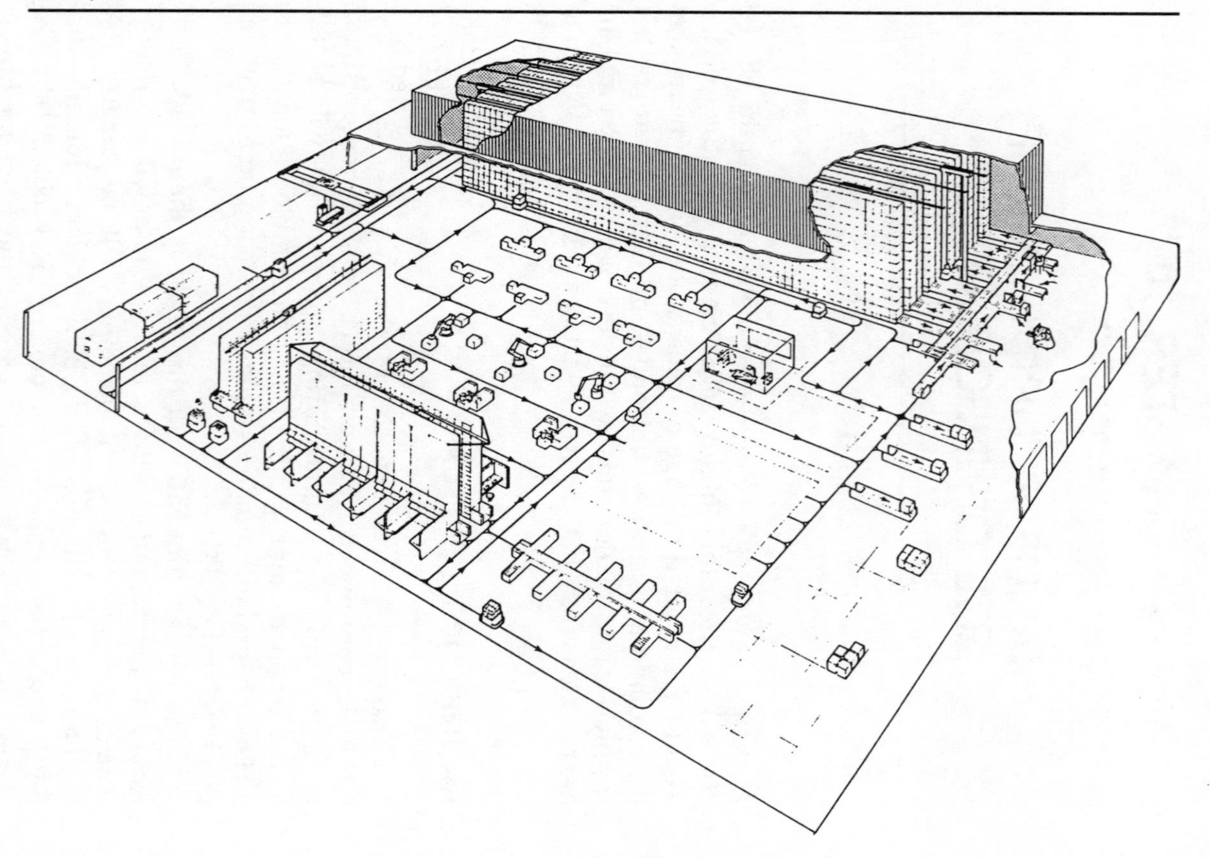

There is no dearth of theories and techniques for achieving the totally automated factory. To consider them all would be impossible; to consider only the best would be presumptuous. So we will look into those that are the most talked and written about: flexible manufacturing systems (FMS), computer-integrated manufacturing (CIM), and optimized production technology (OPT). Many experts give separate definitions to FMS and CIM, but the two are actually almost synonymous. The latter is a logical extension of the former, and that is the way they are handled here.

## FLEXIBLE MANUFACTURING SYSTEMS

Although FMS is not new, it is inseparable from any discussion of the factory of the future. Nor, at the same time, is all computer controlled automated factory equipment properly classified as FMS. Because it is not new—the first systems dating back to to the mid 1960s—definitions abound. Many are good and close to the mark, but the one offered by Young and Green says it best[1]: "A flexible manufacturing system [consists of] a group of CNC [computer numerically controlled] machine tools linked by an automatic handling system, whose operation is integrated by a supervisory computer control. Integral to an FMS is the capability to handle any member of similar families of parts in random order."

In other words, an FMS is a self-contained group of complex machinery—primarily machine tools, robots, and computers—that can perform all the operations necessary in the manufacture of a number of parts, but whose processing requirements are quite similar. Furthermore, the complete operation, including transport from one machine to the next in succession, must be performed automatically under computer control with little or no control from an operator.

---

[1]Clifford Young and Alice Green, "Flexible Manufacturing Systems," *AMA Management Briefing,* 1986, p. 8.

Finally, the sequence of the parts being processed by this system can be random. Differences in parts that are in the same family can be taken controlled by either the central or a satellite computer. The FMS is considered to have evolved from a group technology cell, which strongly suggests a U-line concept where the load/unload functions are at a common location and handle the same, or at least similar, equipment. Figure 10–2 depicts a typical FMS configuration.

## COMPONENTS OF FLEXIBLE MANUFACTURING SYSTEM

Based on its definition, there are at least three distinct components of a flexible manufacturing system: processing work stations—the CNC equipment; material-handling equipment—most likely a robot, automated guided vehicle system (AGVS); and an automatic storage and retrieval system (AS/RS), or

**FIGURE 10–2**
**Flexible Manufacturing System**

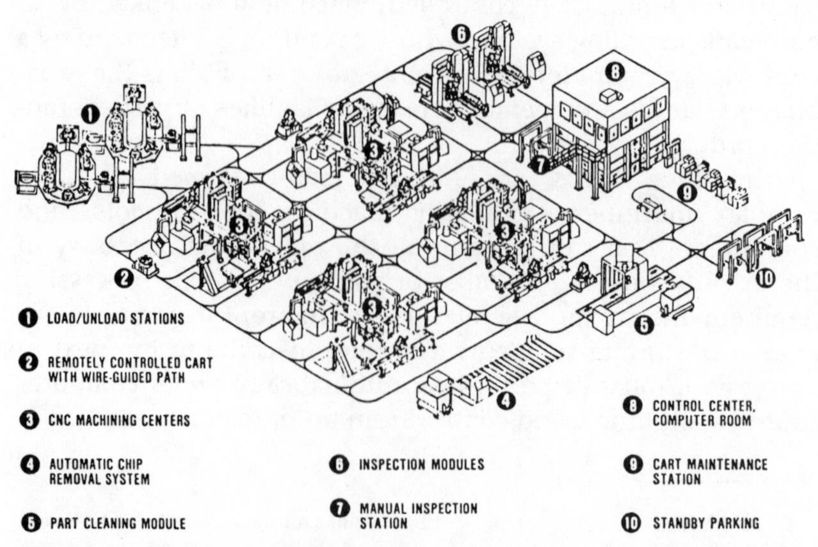


Mikell P. Groover, *Automation, Production Systems and Computer Integrated Manufacturing* (Englewood Cliffs, N.J.: Prentice Hall, 1987), p. 481.

something similar. The final component, central computer control, which directs, commands, and transmits instructions, is really the core of the entire system.

## MATERIAL HANDLING: ROBOTS

It has become increasingly apparent over the past several years that the applicability of robots to a manufacturing-processing environment has been greatly exaggerated. What has been discovered—and is still being discovered—is that the more complex and intricate the task, the greater the probability that the robot will fail to perform satisfactorily. This occurs even when the programming is ostensibly complete and accurate.

It is safe to say that future technology will greatly expand the use of robots, but the tasks they can be expected to perform with current technology are limited. It was not all that many years ago when the use of robots was growing rapidly, only to fall back to a creeping growth more recently.

It is not our purpose to debate the good and bad points of robots. They can perform superbly under a given set of conditions in a given environment, but outside these rather narrow parameters, their use must be seriously questioned.

Industrial robots today are almost exclusively confined to manufacturing, and within manufacturing they are confined almost exclusively to simple and repetitive tasks. Parts requiring exact or perfect positioning would probably not be cost effective in practically all instances. A robot is capable of perfectly performing a work cycle that is exactly the same each repetition, whereas humans are subject to boredom, distraction, and fatigue. Furthermore, a robot is not constrained by time and can work continually through any shift. Among the tasks typically performed are painting, welding, and some simple assembly. The example shown in Figure 10–3 is an arc-welding robot.

Perhaps the greatest application of robots is in material handling, chiefly machine loading and unloading. Furthermore, robots are ideal for handling hot, heavy, or awkward parts and for working in hazardous environments.

**FIGURE 10–3**
**Arc-Welding Robot**

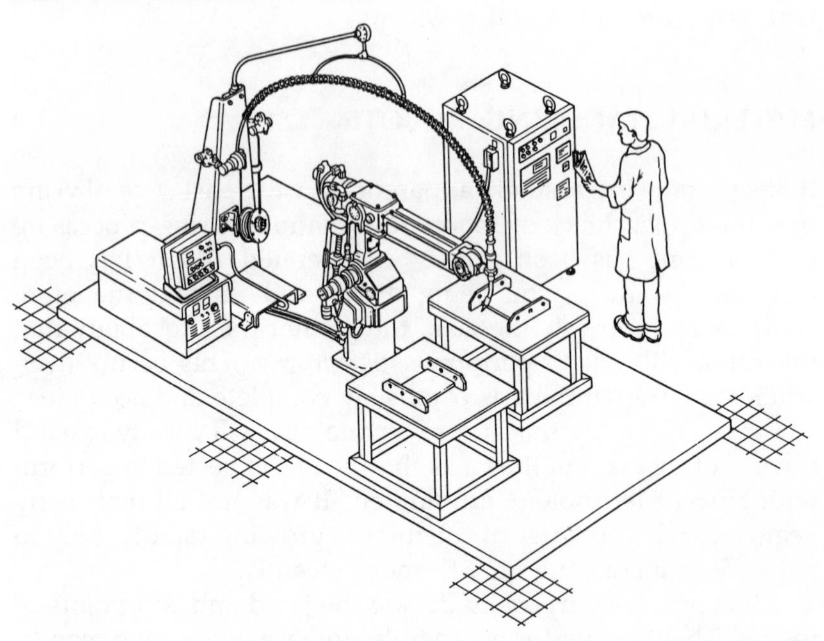

Mikell P. Groover, *Automation, Production Systems and Computer Integrated Manufacturing* (Englewood Cliffs, N.J.: Prentice Hall, 1987), p. 349.

Another main function of robots is automated guidance vehicle systems. Here, too, the route travelled and the range of activities anticipated cannot be too complicated. Beyond this, there is some applicability to inspection activities, but these have to be very elementary. Besides, there are better alternatives available that are not classed as robots. These are covered in the upcoming section.

## PROCESSING MACHINERY

Direct numeric control (DNC) and the newer computer numeric control (CNC) exemplify a marriage between computer technology and machine technology and, as such, represent a giant step into FMS and toward CIM. They were not accidental discoveries nor logical outcomes of technological breakthroughs; they were actively sought after and industriously researched. There were several reasons for this. Whereas numerical-control part-processing represented a clear superiority over conventional means, there were still problems inherent in existing NC systems that motivated machine tool suppliers to look for improvements in the basic systems themselves. The most glaring deficiencies experienced as conventional NCs became more common were these four:

1. Errors in programming: Because these were prepared through human effort, punched tapes for part-programming contained a number of errors. Even though these were always manually verified, mistakes were still bound to exist.
2. Punched tape: In its basic makeup, it is structurally fragile, and through normal wear and tear, processing errors can occur. When Mylar and aluminum tapes came into being, they turned out to be infinitely more durable but, unfortunately, a good deal more costly.
3. Tape reader: Under the technology existing at the time, the electronics that interpreted the instructions contained on the punched tape were considerably advanced but extremely delicate. More downtime was experienced as a result of this particular mechanism alone than all other causes of downtime combined.
4. Controller: The control unit of NC was hard-wired and did not lend itself easily to gradual upgradings or expansion.

There was additional information management sought from the conventional NC that the system was unable to provide. It was therefore an accumulation of all these problems that supplied the impetus for system designers and builders to come up with a better mousetrap. Figure 10–4 shows an NC machining center.

Under direct numeric control a machine or, more typically, a group of machines is controlled by a computer through a direct connection, thus bypassing the need for a tape reader. This improvement made obsolete the least reliable part of the

**FIGURE 10–4**
**Numerical-Control Machining Center**

Photo courtesy of KT-Swasey, Milwaukee, WI.

system and the source of so much downtime. Under this system, a machine or group of machines receives operating instructions directly from a central computer. The demand for instructions is placed by a machine, and the DNC computer communicates them immediately. Because the system is operating in real time, the transmissions of data are instantaneous.

The original change from NC was effected by replacing the tape reader with telecommunication lines. This was a minor and relatively inexpensive change, but as it turned out, the change was only temporary. Newer technology was constantly being developed that resulted in a completely different system. The improvements brought by DNC over its earlier counterpart, however, should be noted. First, it was distinctly more reliable. This improvement alone would have been sufficient. Second, it permitted for the first time, for an entire bank of machinery to be served by a central computer in real time. Third, the computer could be in a remote location, which, because of its physical bulk, was highly desirable.

These apparent advantages become largely illusory when compared to its one overriding disadvantage: What happens when the computer breaks down? The answer: Everything stops!

## COMPUTER NUMERICAL MACHINERY

The advance from NC to DNC to CNC resulted in a significant reduction in the physical size of the equipment and greatly increased computational and storage capabilities. Large mainframes were replaced by minicomputers, which then gave way to today's microcomputers. This advance made it appropriate to locate the computer with the tool it serviced in the same physical plant area or nearby. At the same time, the units themselves were being upgraded. Hard-wiring was replaced by control units. Thus the flexibility of adding to, replacing, or altering the system to one's own specific requirements was tremendously enhanced.

The improvements to part-processing brought about by the development of CNC are being realized today. Some of the more notable improvements over DNC technology are worth listing:

1. The DNC computer sends data and commands to and from a large number of machine tools. CNC computers, on the other hand, control only one machine—or no more than a very small number of them in a single work center.
2. DNC software was developed to convey a huge amount of information. In the days when management information systems were being hailed as the be all and end all, this became part of the total data input. With CNC software, however, any pertinent information is dedicated to the capabilities of a particular machine.
3. Probably most important, the DNC tape had to be fed into the computer each time a part or group of parts was processed. The CNC tape is fed through once and then retained in memory for future requirements.
4. Although CNC tape is dedicated to a particular piece of equipment, it offers a greater variety of features within that environment than did DNC. One example of this is the development of the floppy disc. Changing the programming is now as simple as flipping a switch.
5. Moving to CNC was necessary for advancing to the next sequential step: the control of the total manufacturing system effected through the implementation of either an automatic transfer line (high-volume production) or a flexible manufacturing system (medium-volume production).

## CENTRAL COMPUTER CONTROL

To focus on flexibility and reduced setup times for a sequence of producing activities does not define a flexible manufacturing system. To be an FMS demands integration through central computer control. A random sequence of operations is determined automatically by a central control source, preferably with a minimum of human intervention. Its primary purpose is to respond more quickly to consumer demand, not to reduce manufacturing costs. This important distinction is a departure from traditional management philosophy, and it is a crucial point when the decision of whether to invest in FMS is argued.

FMS directly involves hierarchical levels of computer control: local, area and supervisory. Local, of course, is typically represented by the computer directing the operation of its CNC. Areas are more often minicomputers and, in reality, act as collectors and disseminators of performance data and generally relay instructions down to the individual CNC. These area controllers are sometimes known as satellite controllers.

A recent innovation becoming increasingly popular is the ability of one computer to communicate directly with another through the means of a local-area network (LAN). Any unit connected to the network can talk to any other—across the aisle or across town. The real beauty of this feature lies in the range of devices that can be incorporated into a single network—not only the CNCs themselves, but robots, and bar code readers, data collection terminals, and inspection devices. LANs are especially popular because of their ability to transmit large quantities of data at very high rates of speed.

Current technology is more and more heading in the direction of fiber-optic transmission lines, which offer the ultimate in speed, have more capacity, and are totally unaffected by atmospheric disruptions. There is a downside to the use of fiber optics, however: As you may have guessed, it is extremely costly. Hopefully, further improvements will quickly reduce its cost.

The third level in the computer hierarchy is supervisory control. This links all units together under one central control function. Supervisory computer control denotes a control system in which the computer seeks to optimize some performance objectives for the process, for example, manufacturing levels, maximum yields, or even quality monitoring.

Typically, supervisory control directs the operations of an entire plant. Depending upon the amount of data that must be processed and transmitted, along with other business functions, it must serve other departments, like market analysis, payroll, etc. It may even be necessary to add a fourth level of control. This would be the case where several supervisory computers are tied into the mainframe at the corporate level where they share time with the other corporation activities.

This hierarchical structure and relationship must be understood and its arrangement looked upon as one more step in the direction of CIM.

## FUNCTIONS OF A FLEXIBLE MANUFACTURING SYSTEM

Having already stated what constitutes a flexible manufacturing system, now, perhaps just as importantly, we should state what an FMS is not. The definition of FMS mentioned as one of its main components—probably its key component—the presence of supervisory computer control over manufacturing processes. The reverse of this is not true, however. Supervisory computer control is not the sole property of FMS. A brief rundown of the three classes of manufacturing automation will bear this out:

1. Individual machines are referred to by names like *stand-alone machines* or *islands of automation*. They are individual manufacturing cells called group technology cells and can be looked upon as FMS "starters." They produce a small number and limited variety of unrelated parts, resembling the function of a job shop.
2. Flexible manufacturing systems represent the middle range of manufacturing in relation to quantity: Their volume is too great for individual processes, but not great enough for a transfer line. *Flexible* means that the process can switch back and forth among the grouped items without shutting down each time for the many setups. The FMS represents one of the highest levels of achievement in current automated manufacturing.
3. Transfer lines, also called *dedicated processes*, typify what is commonly termed *mass production*. As in Just-In-Time, the manufacturing processes are set up in a straight-line flow pattern and rarely changed.

It should be noted that in a repetitive, mass-production environment, a measure of flexibility can be provided by means of a programmable logic controller or (PLC). This microcomputer control mechanism is a collection of high-tech gadgetry like relays, timers, and the like, that can be programmed to perform a host of control functions like counting, sequencing, and timing. It was pioneered by General Motors, which is one of the prime users of transfer lines. Figure 10–5 shows what a PLC might look like.

Interestingly enough, the greatest impetus for the acceptance of flexible manufacturing systems was necessity. Increasing world competition demanded products with higher quality and lower costs (and no trade-offs). Due to the aggressiveness of other industrialized nations, survival was (and still is) at stake. And the key to this was flexibility, which led to shorter lead times.

Automation made possible by the computer caused the entire manufacturing process to slowly change, and higher quality, lower costs, and improved delivery capability began to emerge.

**FIGURE 10–5**
**Programmable Logic Controller**

Courtesy of Allen-Bradley, A Rockwell International company.

The goal of FMS was to streamline the processes involved in manufacturing, from the machines that did the actual fabrication to transportation to the storage for tools and fixtures. The driving force, of course, was the customer. Stiffer competition demanded that the manufacturer be able to respond quickly to changes in customer buying patterns while maintaining high-quality standards. This is what is meant by Flexible Manufacturing Systems.

FMS works through the medium of information flow and, as such, is said to be data-driven, as opposed to the physical signals required under conventional batch processing. The importance of this concept must be grasped if one is to gain an understanding of what is involved.

In the material management arena such functions as production scheduling, MRP, and inventory control (including data collected from the shop floor) are all computer-assisted to some degree. They are grouped together under a common concept called computer-aided manufacturing (CAM). Engineering designs parts specifications through a computer-aided design (CAD). The entire process flow, from the design of the product to its completion, is computer-controlled, so the total process is referred to as a CAD/CAM system. The control of the processes as well as the design of the parts themselves, is accomplished through statistical sampling (SPC).

This, in skeletal form, is what is known as computer integrated manufacturing (CIM), referred to at the beginning of this section. As far as state-of-the-art technology is concerned, this is the new kid on the block.

It is the data-driven feature— the flow of information back and forth through the environment in which it works— that allows the various functions and work centers to be coordinated, integrated, and controlled. A typical FMS configuration is presented in Figure 10–6.

## BENEFITS OF A FLEXIBLE MANUFACTURING SYSTEM

1. It can handle a number of different parts so long as the processing requirements are similar.

**FIGURE 10–6**
**Typical FMS Configuration**

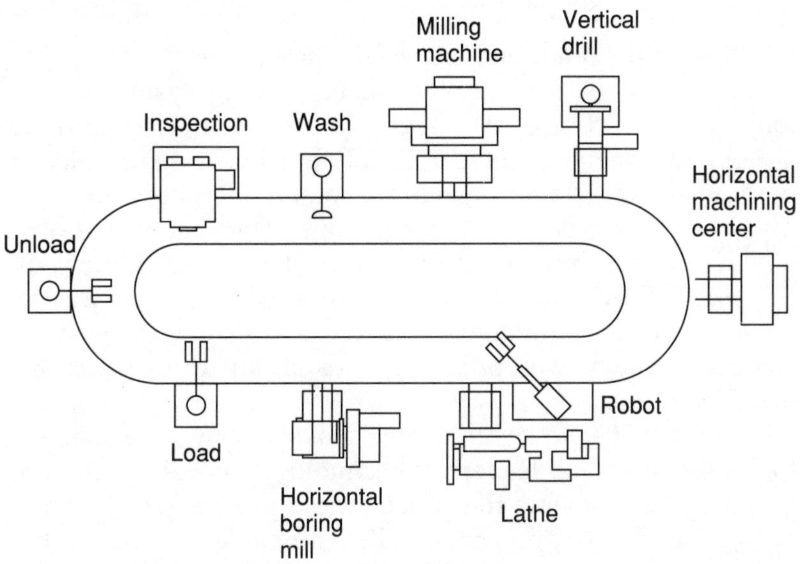


Clifford Young and Alice Green, "Flexible Manufacturing Systems," *AMA Management Briefing*, 1986, p. 9.

2. The various equipment that forms the work center of the FMS will automatically adjust to a difference in part size or configuration upon the direction of the supervisory computer.
3. Automatic transport mechanisms in constant motion keep parts moving into and away from production without the need for any outside intervention.
4. The result of this interaction is greater machine utilization, notwithstanding more frequent changeovers; lower in-process inventories; reduction in direct labor costs, which, in turn, reduces overhead; and a lower reject rate.
5. New members of a family can be added to the product line more easily, probably entailing nothing more than a program modification.
6. As it is characteristically a U-line configuration, an increase or decrease in volume can be handled with relative ease.

7. CAD provides for quick, realizable design change.
8. FMS minimizes the risk of operating in a rapidly changing marketplace.

It may seem strange, especially in view of the foregoing, that the concept of a flexible manufacturing system was so long in the birthing. Although the supposition behind the system had been around for a number of years, FMS did not become a realistic alternative until sometime in the mid 1980s. There were two basic reasons for this. The first seems quite obvious: Competent computer control technology simply did not exist. Furthermore, even when this drawback was removed and the technology did become available, the cost of a complete system was prohibitive for all but the very largest manufacturers.

In the 1980s, however, a move was made by some machine tool suppliers—the leaders in technology improvements and updating—to alleviate the problem to a great extent by advocating a modular approach. Smaller companies, based on their write-off structure, could add individual machines or modules over time, with revenues generated from past improvements. This permitted the entire system to be phased in over an extended period of time, whereas otherwise it would remain beyond their capabilities. God bless capitalism!

The second reason was the recession experienced in the early 1980s, which led to several years of high interest rates. Because these historical "accidents" are now in the past and the need for new technology is still acute, FMS stands on the threshold of a possible tidal surge throughout the 1990s.

## PARTICIPATION IN FLEXIBLE MANUFACTURING

Throughout the brief history of FMS the primary users of automated systems have been limited to large manufacturers, primarily companies involved in some way with transportation: auto makers, truck and heavy-equipment manufacturers, and aerospace.

Initial attempts at flexible automation primarily involved the use of robots for welding, painting, etc., with the aim of reducing direct-labor head counts. This was coupled with computer-aided design and computer-aided engineering (CAE), which greatly reduced the lead time to design, engineer, and tool new-car models. This further reduced manpower costs and did away with, probably forever, the annual shutdown for model changeover.

Meanwhile, changes were taking place in the material composition of automobiles. In keeping with the drift away from the use of ferrous metals, substitutions of lighter metals, like aluminum and magnesium, not to mention plastics, were made. These opportunities were bound to have a significant impact on production technology.

Recent improvements and applications of FMS have included automated-guidance vehicle systems (AGVS) for inplant transportation and bar coding for the purpose of material identification and tracking. Furthermore, computer-aided inspection, by no means a recent development, has continued to add refinements and sophistication to the quality function. Sensory perceivers, especially in those areas where visual inspection is not feasible, are more recent developments. The best known is a probe mechanism positioned on a machine on line, with three-dimensional capabilities, called a coordinate measuring machine (CMM). Because of the highly sensitive contacts and the need for accuracy, this device must be located, positioned, and centered under computer control.

The FMS presented in Figure 10–7 displays a flexible fabricating system for automated sheet-metal processing located in an auto-making installation.

### To Use or Not to Use FMS

It should be obvious by now that FMS is not for everyone. Before considering it as a solution to your future production needs, you must evaluate which and how many parts are intended to be produced. You should keep in mind, too, that you cannot buy an FMS system off the shelf and have your millwright crew bolt it to the floor.

**FIGURE 10–7**
**Flexible Fabricating System for Automated Sheet Metal Processing**

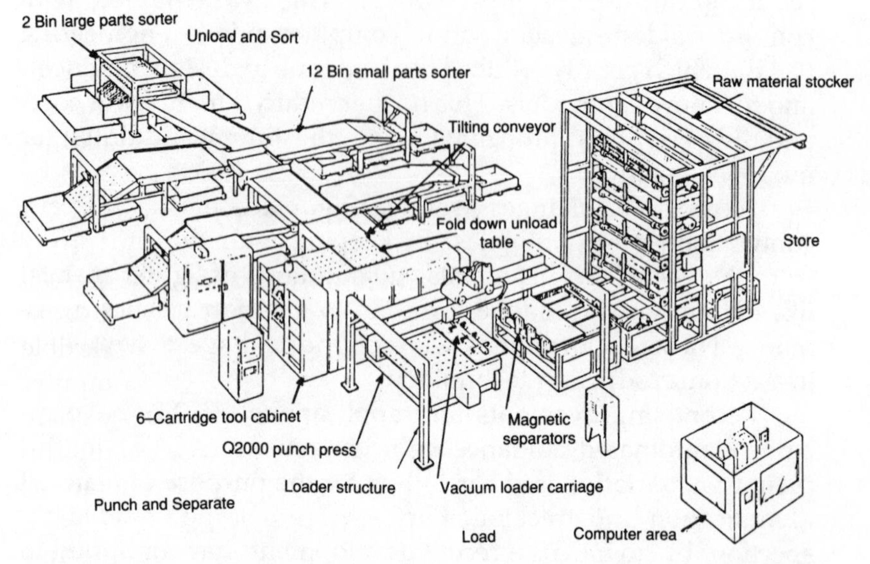


Mikell P. Groover, *Automation Production Systems and Computer Integrated Manufacturing* (Englewood Cliffs, N.J.: Prentice Hall, 1987), p. 482.

Despite these caveats, some manufacturers have bought new capital equipment to make their plants more efficient without deliberating on their long-term needs. One consultant termed this "grabbing at the straws of automation." This conforms to the point made in Chapter 9 regarding trigger-quick solutions to problems: Throw money at them; throw people at them; throw computer programs at them. It calls to mind a talk given many years ago by Ollie Wight: "The Vice-President's New Computer." It was a parody of the children's classic "The Emperor's New Clothes" in which firms purchased computers merely because it was the thing to do and used them only to run their payroll! How far we have progressed in a generation.

Recently FMS installations have tended to be in medium-sized and even smaller firms. Because direct labor costs in these companies generally consume a higher proportion of the production dollar, management's attention to this area commands top priority. Because smaller firms are more capital intensive, those that wish to avail themselves of this newer technology will necessarily adopt a more conservative approach. Moreover, the capability of the larger, more complex systems is completely inappropriate for their more narrowly defined needs. The typical approach is either to use the modular, building-block approach, invest in a smaller FMS, or both. It would be a highly unusual application where a smaller system could not be easily added to as the need arises.

One point bears repeating: An automatic machining cell (for example) or a whole battery of such cells, even interconnected, do not an FMS make. Recalling the definition of flexible manufacturing, it must necessarily include central computer control. The machines may act independently, but they share the same supervisory controls. Additionally, in most applications both the user and the system vendor will cooperate as partners during the entire development phase.

Breaking down the cost of the typical flexible manufacturing system shows the following percentages:[2]

Machines: 50 percent.
Material handling: 10 percent.
Control: 8 percent.
Tools and fixtures: 25 percent.
Services: 7 percent.

It may seem curious that the machinery that actually does the manufacturing only consumes half the system expense. One might expect it to be higher, as it certainly is with less flexible, dedicated systems. Furthermore, the control feature, the system's software, is the least typical of the above breakdown and could actually be much greater than the 8 percent shown.

---

[2]Clifford Young and Alice Green, "Flexible Manufacturing Systems," *AMA Management Briefing*, 1986, p. 20.

## PLANNING THE FMS INVESTMENT

It is important to consider not only the number and types of equipment that will be used in an FMS installation, but also what activities each will perform. Should they be multifunctional, or should they serve a unique purpose? Future additions will need to be planned well ahead of time.

The most important consideration is obviously the actual dollar commitment, which will range from one to several million dollars—or it could be many times that amount, by any standard, a substantial cost. That is why utmost care must be exercised at this crucial stage to systematically draw inputs from each area of the company. A good approach would include some or all of the following steps:

1. Define the manufacturing mission. From page one of this book the basic mission of a manufacturer has been stated unequivocally: Create a customer—which also implies better serving existing ones. This premise has not changed and will not change. What needs to be defined here is how the process will enhance the company's position as it relates to customers' needs, as well as aiding in acquiring more customers. Measuring the impact of the system will likely involve detailed strategic planning.
2. Put together an internal FMS project planning team. As important as the makeup of the planning team and the goals to be fulfilled is the calendar for implementation. This is quite different from the Just-In-Time strategy, where the process of gradual improvement cannot be rushed. Here, unless accompanied by an "accomplished-by-when" schedule, the whole process is bound to move slowly. It is also important that one member of the team be a senior operating officer, which will ensure cooperation and visibility.

   Equally important is the inclusion on the team of the technicians who will operate the system. Their input can be invaluable, especially when expanding existing systems.

3. Get to know and understand the technology involved. This could just as easily be the first step as the third one. Unless you are adding to an existing system, you must be able to grasp fully what you are letting yourself in for. Only you, as an individual or a company, can make that determination. Furthermore, be advised: *FMS is not a solution to current production problems!*

   Self-indoctrination would be a logical first step here. Attending conferences, perusing current periodicals, visiting ongoing installation, and learning from the technology suppliers are some of the ways to accomplish this. This would be followed by liberal doses of educating your internal work force and drawing from their considerable experience.
4. Conduct a preliminary evaluation. If the progression thus far has been steady and positive, you should determine the following: Does the mix of parts produced in your plant make them likely candidates for FMS? How many families of items are there, and what are their respective volumes? This is where Engineering can provide valuable insights. And Marketing may be able to project what the future holds for existing products and the adaptability of new product lines to the proposed technology.

   Having progressed to this point, you are now ready for the final step. It is time to make your decision. Your detailed analysis reveals whether a new automated system makes good economic sense. If it is a go, preliminary contacts with several different vendors is in order. Their ideas and suggestions can only be helpful.
5. Prepare a request for quotation. The most important thing at this stage is to ensure that what you are proposing to buy and what the vendor is proposing to sell are one and the same. The relationship between you and your vendor is a practical application of what was covered under Just-In-Time. The freer the flow of information between the parties, the more informed you will be and the better able to make an intelligent choice when the time comes.

6. Evaluating the individual vendor. You are the most informed about your operation, and so it follows that you must determine which vendor, if any, best fits your list of requirements. Try not to yield to the temptation of buying the least expensive system for that reason alone. If you have been diligent throughout the planning process, you will be armed with the ammunition to make the correct choice.

## INVESTMENT ANALYSIS AND JUSTIFICATION

Somewhere in the evaluation process you must justify the entire system as an investment. If a commitment seems likely to be made regarding FMS—or even more broadly, regarding CIM—how do you calculate that it is a wise decision from an economic standpoint? Traditionally this has been done through a cost/benefit analysis.

A cost/benefit analysis is certainly valid if you are investigating the financial feasibility of purchasing a new piece of capital equipment. A payback period of cost savings can be calculated with great accuracy, but this is not what you are dealing with here. This new technology is going to revolutionize your corporate modus operandi for a very long time to come. And there are forces at work here that do not fit easily into either cost or benefit analysis.

One question you might well ask yourself is, "If not FMS, then what?" (This assumes that standing still is not an acceptable alternative.) What else is out there that will keep your factory efficient and competitive? This is more a strategic decision than an economic justification. If the decision is to pursue FMS, you have made an investment in the future, and that future does not necessarily have a short horizon.

There are costs as well as benefits that can be defined with relative accuracy. Obviously the cost of a landed system is a given because this constitutes the hardware, programs, and controls you bought. Further, there will undoubtedly also be some out-of-pocket costs for additional software and, if the situation warrants, other pieces of hardware. Training costs to acquire new skills must be considered, as well as the potential need to add a few trained technicians to the payroll.

On the benefit side, the saving in direct labor costs is easily computed. There are indirect cost savings as well, but these are a little harder to get a handle on. Included in this category would be such things as in-process inspection, transportation, shop floor control, and so on. The cost savings in dollars and cents are real; they just aren't apparent or as easily quantifiable.

As you progress down the list of considerations, the ability to assess a given cost becomes increasingly difficult, especially when it involves a longer time frame. For example, how would you compute the costs and benefits of fine-tuning the system in order to realize its full potential? As a pioneer in FMS once remarked, "Justification should be based on your convictions, not your accounting."[3]

## COMPUTER-INTEGRATED MANUFACTURING

Implementing a working flexible manufacturing system requires, in order, proficiency in machine tools, in automated handling systems, and in the network of computers. This puts it squarely on the path toward CIM. Within this network, it becomes increasingly important to understand how the different automation systems interact with one another and the common data needs of each device. As things stand now, CIM represents the ultimate in information processing.

Computer-integrated manufacturing brings together and merges all the information-processing functions necessary to support the design and manufacture of a product, as well as the myriad planning and control functions needed for reinforcement and the conventional business operations of the manufacturer. A complete CIM system is shown in Figure 10–8.

This meshes well with the definition of CIM given in the *APICS Dictionary*: "The application of a computer to bridge and connect various computerized systems and connect them to a

[3]Young and Green, p. 52.

**FIGURE 10–8**
**Complete CIM System**

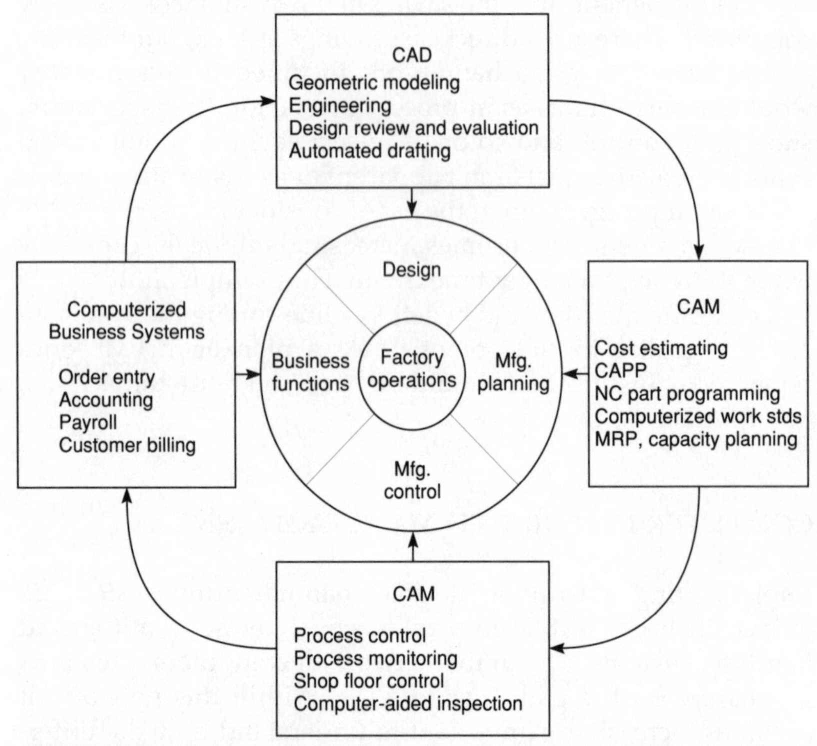


Mikell P. Groover, *Automation, Production Systems and Computer Integrated Manufacturing* (Englewood Cliffs, N.J.: Prentice Hall, 1987), p. 722.

coherent, integrated whole. For example, budgets, CAD/CAM, process controls, Group Technology systems, MRP II, financial reporting systems, etc. would be linked and interfaced."[4] Full implementation of CIM results in the automation of the information flow through every aspect of the company's organization.[5]

Flexible manufacturing systems, even those encompassing CIM, do not provide for the manufacture of products more

[4]*APICS Dictionary,* p. 6.

[5]Mikell P. Groover, *Automation, Production Systems and Computer Integrated Manufacturing* (Englewood Cliffs, N.J.: Prentice Hall, 1987), p. 723.

cheaply. Their chief benefit is to allow for things to be done that were previously not possible. That is why the standard cost/benefit analysis (return on investment + payback period) is not appropriate when considering adoption of this newer technology. If FMS and CIM are truly the new wave of manufacturing and are part of the factory of the future, what other changes are they bringing with them?

1. Shortened product life cycles. This has nothing to do with wear or "planned obsolescence." It is totally due to the pace of new technology, not to mention competitive pressure, which will render existing products effectively obsolete.
2. Emphasis on quality and product reliability. The quality/price trade-off will have completely disappeared because quality will replace price as the principle determinant.
3. Downsizing. Everything from the size of the products to the size of the workforce will be decreased. Reducing the size of end products through the growing use of microelectronics will tend to lower the cost of getting them to the market. Technicians and generalists will more and more replace specialists. This, plus the need for capital, will keep companies lean.
4. Reduced inventories. The age-old sentiment that success is created by high inventory levels will disappear. Competition and cash flow needs will tend to keep them low.
5. Greater evidence of Just-In-Time manufacturing. JIT will grow as more customers come to accept quicker deliveries of smaller lot sizes. This will tend to keep inventory levels in line also.
6. Greater emphasis on the computer. This almost goes without saying. Interest in computer-integrated manufacturing will continue to grow, and CAD/CAM users will become more numerous as smaller firms begin to take advantage of them. The popularity of FMS, PLC, and microcomputers will spread as the age of information takes hold. Management's focus will switch from a database to a knowledge base.

## OPTIMIZED PRODUCTION TECHNOLOGY

The proprietary system called optimized production technology (OPT) has received quite a bit of notoriety over the past several years. The creation and development of OPT is credited to the popular author and lecturer, Eliyahu Goldratt. It is variously referred to as a manufacturing philosophy and as an enhancement to basic manufacturing systems.

The system is run on a trademarked OPT software package that is composed of four modules: buildnet, serve, split, and OPT, or scheduling. The purpose of the modules, in succession, is to process manufacturing data inputs like bills of material and routings to form a product network. Then, using the information in the product network, the system works backwards from order due dates in an attempt to identify bottleneck work stations.

Also incorporated into the system modules are rough-cut capacity-planning routines (load profiles) that compute the average utilization of each manufacturing process. These are then sorted into bottleneck and nonbottleneck resources. The fourth module, OPT, utilizes an algorithm (also proprietary) that attempts to arrive at the "optimum" scheduling solution using finite-capacity-scheduling logic. This logic would normally include capacity and manpower constraints, maximum inventory, batch sizes, and other information collected during the product networking stage and categorized under the resource description.

The main contribution of OPT is that it handles the rough-cut capacity-planning function in a new way. It maintains that critical resources are the constraining condition in any shop. This recognizes the fact that bottleneck operations are the constraints and must be the starting point in order to construct a realistic Master Production Schedule. Once bottleneck resources have been identified and scheduled using the finite-loading algorithm, the balance of the shop, or nonbottleneck work stations, can be scheduled using standard MRP procedures. OPT cannot operate using such procedures because, even though it proposes lot-for-lot run sizes, it makes use of safety stocks, safety lead time, and order splitting as standard techniques.

What OPT tries to do is maximize the uptime of any bottleneck resource. It therefore entertains no preconceived notions regarding run sizes. This is in direct opposition to the finite-loading procedures typical of batch processing. OPT says that batches can be the same size or just as well different sizes. In other words, OPT focuses on the bottleneck resource and not on the product being processed. It is the sequence that is scheduled, not the order quantity.

Another contribution of OPT is offered by the system software: the way in which it evaluates information inputs in order to identify errors that may happen during data processing. The typical error under OPT is scheduling a bottleneck resource in excess of 100 percent of its capacity. As a matter of fact, even loading it as high as 100 percent is disallowed.

The theory behind OPT is as interesting as it is unique. It sets forth that an hour lost at a bottleneck work station is an hour worth of output lost to the entire factory, while an hour lost at a nonbottleneck operation carries no real cost. This may not be wholly true, but that does nothing to diminish the impact of the message it delivers. In the meantime, however, there is still the question of cost—whether it is real or not is merely a matter of interpretation.

OPT recognizes that the physical resource, whatever its classification, carries with it some definable cost. To this must be added a percentage of overhead and a proportional share of energy the resource consumes. Finally, there is a workforce that must be paid, regardless of whether they are involved in some related work activity. Therefore, there are fixed costs associated with either situation, but the hour lost at nonbottleneck resources carries with it no additional cost because it has already absorbed the fixed portion. The hour lost at the bottleneck work station, however, is lost forever and can never be made up.

The overall goal of any operation is to make money. This is accomplished by maximizing throughput. Throughput is constrained by bottleneck work stations, so the focus of attention should be on maximizing output through the best mix of capacity utilization. Only throughput can become an account receivable.

To achieve the objective of maximizing throughput, OPT provides a workable alternative to priority loading or scheduling, as would be the case under an input/output method. This doesn't mean that OPT supplants Just-In-Time. Rather it seems to complement it. Under either system the concentration of effort remains fixed on value added and not on cost incurred. A flowchart of the OPT system is shown in Figure 10–9.

## SUMMARY

In the main, the difference between group technology cells and flexible manufacturing systems is that the latter involve the application of computer-based technology to processing and material-handling activities. In other words, operators in GT cells perform the functions of loading/unloading, equipment changeover, assembly, and inspection, whereas under FMS these are all automated. Under Just-In-Time this is what logic prescribes. As the operation of a GT cell matures, further improvements can only come through automation.

Several factors that bode well for the future of FMS are a direct reflection of what stifled its growth in the early years. At that time computer control technology was not sufficiently advanced, which is no longer the case. The earlier systems were highly specialized, with narrow and specific objectives, but they proved successful almost from the start. This prompted others to research the possibility of adapting this new form of automation to their own operations.

The last factor was the ultrahigh cost of the new technology. Going the modular way only made it less painful. As history has shown many times, as demand rises, competition increases and costs will tend to moderate.

Finally, the question might be asked, "Which system should I use?" There are a vast number of techniques available to solve any number of problems. The ones discussed here are not necessarily the best, only the most popular. If you must choose between MRP, JIT, and OPT or automated versus nonautomated equipment, what is the best way to go? That depends on how you answer two questions. The first is simple; the other is a good deal more complicated.

**FIGURE 10–9**
**OPT Flowchart**

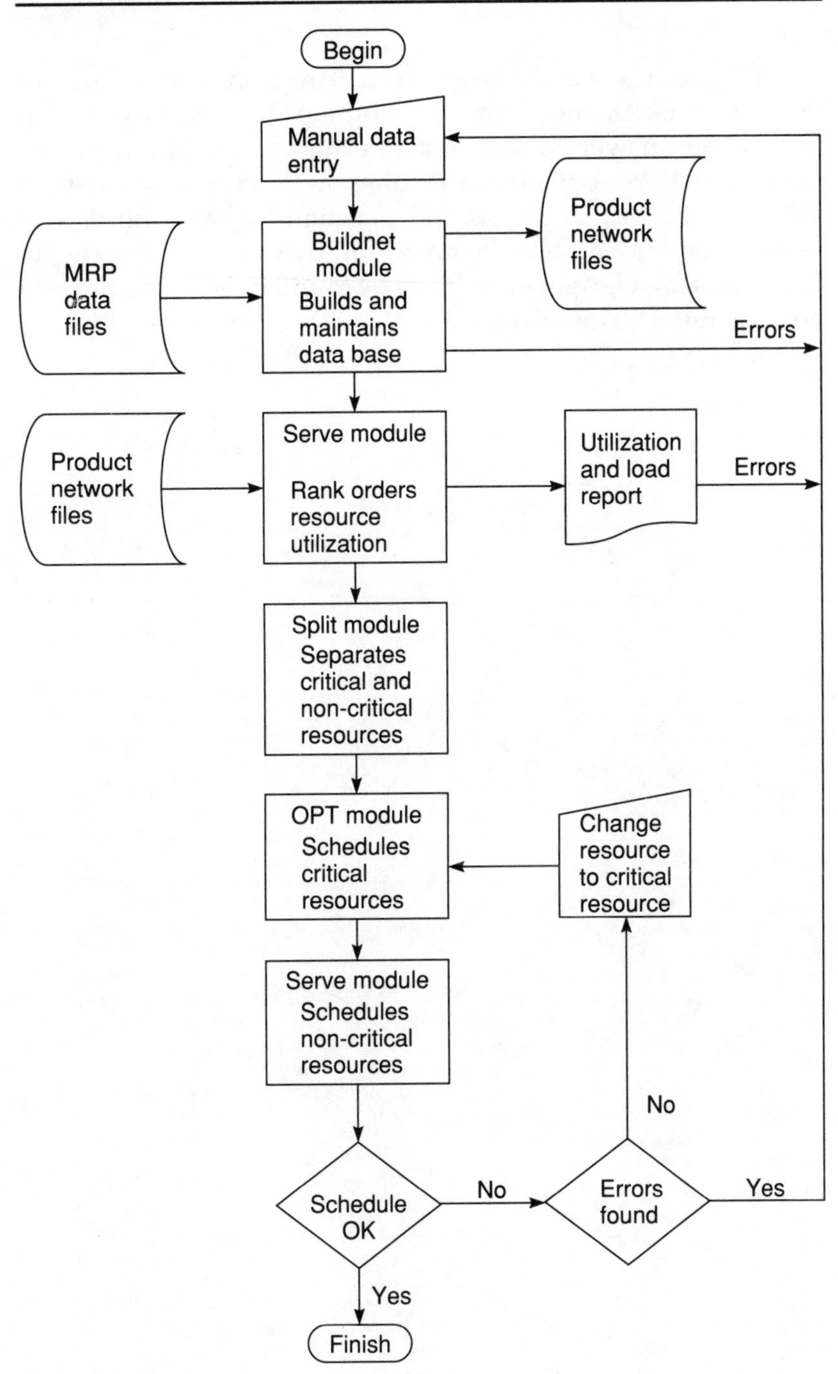


Thomas E. Vollmann, "OPT is an Enhancement to MRP II," *Production and Inventory Management*, Second Quarter 1986, p 16.

Two things we know are true: The computer is here to stay, and technology can do nothing but improve. That makes the answer to the first question—automation versus nonautomation—obvious. To not move forward is to stand still. The answer to the second question—what controls will serve us best?—can only be answered by you. And the clue to the correct answer can most likely be found in how you defined your manufacturing mission.

# CHAPTER 11

# MANAGING IN THE AGE OF INFORMATION

Many writers have delved into the implications of automated manufacturing for the makeup of personnel in the factory of the future. Obviously it would most affect unskilled and semiskilled workers. Many who have researched the subject have concluded that the factory of the future will be peopled only with skilled help, in three categories: technicians and engineers—machine people; programmers—computer people; and managers—information people.

While the signs for other workers are certainly far from encouraging, it is still not the time to throw in the towel. The history of our free-enterprise economy shows that somehow events always seem to occur to take care of, or at least alleviate, any massive displacement of workers. As everyday life becomes increasingly high tech, acquiring a certain level of skills is mandatory just to stay even. But there are enough indications that everything is not total gloom and doom.

## OPPORTUNITIES WILL STILL EXIST

Recall the last technological revolution, just 30 years ago, heralding the age of computers. At that time the major "advantage" was seen as the number of clerical workers the computer would replace. As it turned out, it didn't replace anyone; it merely generated a lot more data and reports for the existing staff to read, digest, react upon, and disseminate. In fact, the lasting effects for workers have been positive. The more computers there are in our everyday lives, the more personnel will be required.

Nevertheless, it is the expressed intent of automation to reduce direct heads, and this it is doing. But who is to say that this is all bad? Certainly it may be a personal catastrophe for those who are being displaced. But over time, will this not free up capital for expansion that will create more jobs? It's true that these jobs will be mostly for skilled technicians and not for those laid off previously, but one social objective is not to experience a net loss in job opportunities.

How about the service industries that will be necessary to support this blossoming technology? These are but a few of the opportunities that could result from the advance of automation. Nevertheless, the crux of the situation is this: Social implications aside, the new technology must be accepted as the price the United States must pay to remain competitive in the world. Anything else would be national folly.

Another indication that all is not negative lies in the fact that this technology is not all-encompassing, nor are all the changes likely to occur everywhere at the same time and at the same pace. Experts agree that FMS is applicable to no more than 35 percent of industry, and its first cousin, transfer lines, a whole lot less than that. While this does not mean that the rest of society will be largely unaffected by these advances, the change is mostly one of degree. Moreover, there are segments of industry whose method of operation does not lend itself well to automation. Among them many types of plastic molding and job shop operations. From all indications, these segments are not only healthy but growing.

FMS is most applicable to firms engaged in metal-working and machining products, so there are more ways of arriving at computer-integrated manufacturing than just FMS. After all, CAD and CAM are applicable to all forms of manufacturing, not just a select few.

Finally, there is nothing inherent in automated manufacturing that requires a complete overhaul of our commercial structure. Business will be carried on in much the same fashion as it has in the past. Customers must still be canvassed and accounts balanced. Suppliers must be paid, customers must be billed, payroll must be run. Controls of various kinds must continue to be maintained, and last but not least, the organization will always have to know if it is succeeding, is just breaking even, or is in a "projected negative-cash-flow position."

To argue that business is not changing is ridiculous. It could well be that, over time, the very fabric of manufacturing will change, but it will be the result of gradual improvements, evolution, rather than a cataclysmic event. Therefore, it becomes the responsibility of every executive and every long-range planner to recognize the problems that will accompany the opportunities these changes will bring.

## DOWNSIZING

Over the past decade it has increasingly been the case that companies have not filled many staff positions that were traditionally part of a growing corporation. This has occurred for several reasons.

First, intense global competition has forced most corporations to lay off all but "essential" employees, with the job responsibilities distributed to the remaining personnel. A good example of this can be found in the auto industry during the recession of the early 1980s, when scores of highly paid executives were laid off in Detroit. Few of them were ever likely to be called back, but those few that were received greatly reduced salaries and maybe different job descriptions.

Second, a new kind of contract service has emerged that will, for a fee, take care of many of the services formerly the domain of the employer, the theory being that these contracted firms can perform the services at a lower cost than can the individual company simply because of economies of scale. Several of the common services they handle are payroll, pension plans, and insurance administration. This trend shows no signs of abating anytime soon. Some companies have even turned over to a contract service the total personnel administration of an entire factory.

The result of all these policy changes has come to be known as downsizing. Downsizing should not be confused with layoffs, which carry with them the notion that there will again be employment when the cause for the period of adjustment is removed. Although downsizing produces exactly the same result, getting rid of people, it comes from an entirely different perspective. There is no attempt to level the work force in response to the vagaries of the economic cycle. The goal is to run lean at all times, with no open-ended expectation that the company may someday be "upsized." Downsizing is a direct management decision, and it is accomplished by natural attrition and early retirements, in addition to actual terminations. Some companies believe that the word *downsizing* carries with it too harsh a connotation, so they have termed the process *rightsizing*. It may sound better, but it hurts those affected just the same.

## THE MANAGER

What qualifies a person to be a manager? And what kind of people will he or she manage? It is well recognized that managers will necessarily be generalists. To be successful, they will have to have more than a passing acquaintenceship with the operating departments outside the sphere of their training and experience; they must be able to comprehend the complete manufacturing enterprise.

## THE CIRM PROGRAM

Ever since its inception in the late 1950s, the American Production and Inventory Control Society (APICS) has been the bellwether in the area of manufacturing theory. It has been at the forefront as manufacturing's staunchest crusader, and it has also been the organization most responsible for raising the status of the manufacturing professional. It is the latter accomplishment that has provided APICS with its greatest visibility.

Probably the single most important reason for its success is the training APICS provides on an ongoing basis to all practitioners, regardless of the level of their expertise. For the past 20 years APICS has awarded certification, as Certified Production and Inventory Manager (CPIM), to professionals who have achieved a given level of attainment.

Recently, however, APICS's vision has shifted to some degree, as it has looked into the future and attempted to visualize the manager who will be guiding the manufacturing organization in the next century. It has been their conclusion that expertise in production and inventory management is no longer sufficient and that a better balance of skills will be necessary.

APICS has proposed that the new manager be a Certified Integrated-Resource Manager (CIRM), and it is setting out a course of study leading to this. Not only must a manager be proficient in the management of materials, he or she must also, as it becomes necessary, cross functional lines and be equally adept in handling a product, accounting, or quality problem. Furthermore, conventional functional lines will have to be considerably broadened or, even better, will disappear altogether. Just as one of the benefits of FMS is the ability to change quickly, so must the manager be flexible and decisive. Management by committee will become a relic of the past.

The CIRM program is presented in a series of five modules, each with an examination to test the examinee's knowledge and understanding of that functional area. Successful completion of all five examinations will earn the individual the CIRM award.

At present the program is still early in its development and implementation stage. The proposed texts, reprints, and other program review material (bibliography, dictionary, etc.) have yet to be compiled into a complete certification study guide. The completed program is due to be in place by the spring of 1993. Even then, constant revisions will be necessary in order to stay current. The five functional areas and their subsystems are shown in Figure 11–1.

The factory of the future will promote the idea of the multifunctional manager. But how will the manager manage? Probably in very much the same way as he or she is managing today. A manager will probably need to possess better communication skills, and may well have a better-educated, better-motivated workforce—as well as a lot smaller one. Yet virtually all the major changes will be going on behind the factory wall,

**FIGURE 11–1**
**CIRM Modules**

Customers and products
- Marketing and sales
- Field service
- Product design and development

Logistics
- Production and inventory control
- Procurement
- Distribution

Manufacturing processes
- Facilities management
- Process design and development
- Manufacturing

Support functions
- Information systems
- Human resources
- Quality management
- Accounting and finance

Integrated manufacturing management
- Interrelationships of all functional areas throughout an enterprise

Source: *APICS CIRM Certification Study Guide* (City, XX.: APICS, Year).

where many direct labor heads will be replaced by a handful of technicians and programmers.

In front of the factory wall, life will go on pretty much as usual. The manager will have a lot fewer levels to sift through to nail down a problem, but that's pretty much it. All the business functions that cause the firm to operate effectively will still be in place. Moreover, all the information that is being generated must be read and acted upon by someone. It's easy to comprehend the concept of a dark factory, but a dark office is extremely unlikely.

## SUMMARY

If the age of information is not here yet, it is certainly creeping up loudly. It is difficult to visualize exactly what will happen in the future, but there is no doubt that technological innovation and information generation will continue unabated.

Unfortunately the lives of many people are going to experience upheaval and dislocation. The worst part is that it can't be helped. Automation can no more be stopped than can a winter snowstorm. To not move forward is to fall behind, and it can only be expected that the welfare of the nation as a whole will benefit.

## CONCLUSION

*Common Sense Manufacturing* is intended as a handy reference tool to solve some individual problems. Obviously the entire gamut of controls and techniques has not been presented. We have attempted to separate the good techniques from the not-so-good and explain how a given concept is supposed to operate and in what kind of environment.

This book is *not* intended to suggest what system you should use in your operations, for they are your own responsibility. Only *you* can determine what will be effective.

We have compiled this book for those who desire to put in place an MRP II system but have found their current practices falling somewhat short. It should be noted that you can no more acquire an MRP II system than you can buy and put in FMS. The system must be built from within by the people who will operate it, with outside help used to get over the rough spots. MRP will be tailored to the unique set of circumstances in your company.

Throughout the many months of research and compilation of information for this book, I have kept a weather eye on the economic scene, especially U.S. manufacturing. Things are happening faster than anticipated, and the signs are mostly positive. We are actually winning the war. And we are doing it in the typically American fashion: We have overcome adversity through sheer determination and hard work.

# BIBLIOGRAPHY

A complete bibliography is available from the American Production and Inventory Control Society (APICS), 500 West Annandale Road, Falls Church, Va. 22046–4274. This periodically published, comprehensive compilation covers not only the field of production and inventory control but the entire spectrum of manufacturing, including ancillary activities and support systems.

The following list of recommended titles is certainly less extensive than the APICS bibliography, but it does contain reference material that should prove to be useful to the student or professional requiring further information on production and inventory control.

American Production and Inventory Control Society. *APICS Dictionary.* 6th ed. Falls Church, Va.: APICS, 1987.

This dictionary clarifies confusing words and phrases and provides a universal meaning for new concepts as they evolve.

Deming, W. Edwards. *Out of the Crises.* Cambridge, Mass.: MIT Press, 1982.

The author's style, eminently readable, is truly unique. This is 500 pages of common sense. A shorter book, *The Deming Management Method,* is also recommended reading.

Fogarty, Donald W., and Thomas R. Hoffman. *Production and Inventory Management.* Cincinnati, Ohio: South-Western Publishing Co., 1983.

This teaching tool is one of the basic building blocks for APICS certification. It is heavily involved in the technology of systems.

Ford Motor Company. *SPC Manual.*

For information on this title, contact: P. T. Jessup, Corporate Quality Education Center, Ford Motor Co., Room 524 WHQ, P.O. Box 1899, Dearborn, Mich. 48121–1899.

Groover, Mikell P. *Automation, Production Systems, and Computer Integrated Manufacturing*. Englewood Cliffs, N.J.: Prentice Hall, 1987.

Probably the first book totally dedicated to automated production systems and how they work. Readable and comprehensible, with enough technical background and features to satisfy engineers who are interested in investigating automation.

Hall, Robert W. *Zero Inventories*. Homewood, Ill.: Dow Jones-Irwin, 1983.

In collaboration with APICS, Hall has examined how Japan has revolutionized manufacturing and how its practices could be adapted for American industries. This book remains the primary reference for Just-In-Time.

McGuire, Kenneth J. *Just-in-Time Manufacturing*. MGI Management Institute, 1988.

Another APICS-sponsored textbook by a recognized authority on Just-In-Time and its transfer from Japan to U.S. manufacturing facilities. It is both readable and instructive.

Orlicky, Joseph. *Material Requirements Planning: The New Way of Life in Production and Inventory Management*. New York: McGraw-Hill, 1975.

In the author's own words, this is a book on MRP from A to Z. It is generally considered to be the primary reference source for system construction, implementation, and operation. A note of caution, however: The book does not discuss MRP II, which was not developed until several years later.

Plossl, G. W. *Production and Inventory Control: Principles and Techniques*. Englewood Cliffs, N.J.: Prentice Hall, 1985.

This is an updated edition of the original work coauthored with the late Oliver Wight (see next entry).

Plossl, G. W., and Oliver W. Wight. *Production and Inventory Control: Principles and Techniques*. Englewood Cliffs, N.J.: Prentice Hall, 1967.

This publication stands as a monument to its two authors, for it brought production and inventory control out of the closet and gave it meaning and respectability. Because it preceded the dawn of the MRP crusade and the computer revolution, it is now seriously outdated, but it still remains a favorite of those who remember the good ol' days.

Vollmann, Thomas E.; William C. Berry; and Clay D. Whybark. *Manufacturing Planning and Control Systems*. Homewood, Ill.: Dow Jones-Irwin, 1988.

A thorough, intensive treatment of the state-of-the-art in manufacturing practices for progressing through the 1990s. Written as a textbook or practitioner's manual, this book examines in depth the concepts of optimized production technology (OPT) and flexible manufacturing systems (FMS) as these impact on today's global markets.

Wight, Oliver W. *Manufacturing Resource Planning: MRP II, Unlocking America's Productivity Potential*. Essex Junction, Vt.: Oliver Wight, Ltd. Publications, 1984.

Sums up the author's many years of experience in manufacturing management. He makes the point that MRP II encompasses a great deal more than just what is happening on the shop floor.

Wight, Oliver W. *Production and Inventory Management in the Computer Age*. Boston, Mass.: Cahners Books, 1974.

Although this book preceded by about a year Joe Orlicky's *Material Requirements Planning*, the topics of the two books are quite similar. But here the author gives added emphasis to the importance of priority planning and control.

Young, Clifford, and Alice Green. *Flexible Manufacturing Systems*. (AMA Briefing Report) New York: American Management Association, 1986.

A good nuts-and-bolts resource manual on what is turning out to be one of the emerging technologies of the 1990s.

# APPENDIXES

# APPENDIX A

## THE JAPANESE CONNECTION

I have never been to Japan. Someday—God willing and the creeks don't rise—I will go there. But when I do, it will be as a tourist and not as a member of some study group. Nothing so tries my patience as the incessant stream of people flocking to Japan to confirm what they already "know": that Japanese industry has attained excellence in every manufacturing performance measure there is, while here in the United States we plod along, somewhere in the bottom third of mediocrity. This is, in a word, *hogwash*! Nothing could be further from the truth. These people remind me of the advertising agency that wanted to say that three out of four doctors recommended their product, and the first job for the agency was to go out and find the three doctors.

Don't get me wrong. I don't mean to suggest that the Japanese haven't accomplished what they are being lauded for nor that they don't deserve the recognition and respect they are receiving. They certainly do. They have pioneered the most remarkable manufacturing breakthroughs and have done so in convincing fashion. It would be extremely naive and ostrichlike to pretend otherwise. Nor is this intended to be an indictment of the Japanese people. If they are guilty of anything, it is grasping and making use of the opportunities that have come their way.

The point of the issue is this. The findings of the study groups and the plethora of articles describing Japanese superiority notwithstanding, no one seems to have taken a step back, looked hard at the situation, and asked the obvious question: Does it seem reasonable, given the might of American enterprise and the relative weakness of the Japanese system, that through improved manufacturing practices alone, the Japanese could first equal and then surpass the productive efforts of some of our basic industries? And furthermore, accomplish all this in less than a generation? It hardly seems possible that it could be true, and it's not. Moreover, the evidence supporting this contention is too overwhelming to be ignored.

To have come so far in so short a time means that Japan had to have some help. This help came in the form of two events of gigantic proportions that almost completely overshadow the remarkable strides that Japan made in manufacturing. The first event is unlikely to ever be repeated; the second seems unlikely ever to stop. The first event began in 1977. The White House was occupied by an unbelievably incompetent, born-again peanut farmer and his equally incompetent staff of advisors. This artless group managed to accomplish, in less than two years of ineptitude, to take a healthy economy and turn it into a shambles. Inflation, interest rates, and the price of gold reached all-time record highs while the value of the dollar, not to mention the confidence of our allies, was at a record low.

The decision was made to sell off some of our gold reserves to prop up the dollar. However, instead of parceling out the gold in small lots to slow the devaluation of the dollar in order to stabilize it, practically the entire reserves of our gold bouillon were thrown out onto the market all at once. The far-reaching effects of this little stunt were cataclysmic. What it did was jump-start the value of the dollar and catapult it into a completely new orbit. When it began to rebound, instead of settling down to its own natural level, it shot right past and kept on going. When it finally reached its zenith, it was the costliest money circulating in the industrial world.

Dollars then began to disappear. Because they were so valuable, they were horded by foreign-exchange speculators who used them to buy up marks or yen or whatever at bargain basement prices. Other investors and importers who managed to hang onto their dollars were certainly not about to buy U.S.-made products at prices grossly overvalued in relation to their own.

On the other side of the coin, the impact on the domestic economy was devastating. Unable to sell abroad and unprotected from competition here at home, the American productive machinery practically ground to a halt. The flood of Japanese products coming into the country reached epic proportions and caused some temporary restrictions to be imposed. Of course by then the damage had already been done.

As the value of the dollar began to recede and seek its own natural level, foreign-exchange investors unloaded what they had purchased for high-priced dollars and reaped huge profits. In addition, as a result of this debacle, the Japanese reinvested their profits right here, buying up still more of our productive capacity. Finally, one fact stands out as clear testimony to this misadventure. The entire chain of events, from beginning to end, had absolutely nothing to do with manufacturing efficiency.

In retrospect, it is easy to see that something had to be done to prop up the price of the dollar to prevent it from possibly becoming permanently devalued. But this was too high a price to pay. The real worth of a dollar is made up of two things: The gold or other form of hard currency backing it and the confidence placed in it by those who use it as a medium for trading goods and services. Because the Carter administration was so ineffective, confidence was badly shaken and this brought about the problems I have described. Because the United States has since learned not to elect any more Carters, this problem is not likely to recur.

However, it is the second event that is truly insidious and threatening: our trade policy with Japan, which is positively inexplicable. If I have ever seen heads I win, tails you lose proposition, this one is dictionary perfect. Part of the reason it is so difficult to explain is because it has been going on for so long, ever since the 1950s and it shows ominous signs of continuing forever. Every time I pick up a newspaper where a headline screams "President Signs New Trade Pact with Japan" I cringe. I know that the article will either say that we have further opened up our markets to their products with no reciprocity on their part or simply re-air the stale old arguments for maintaining the status quo.

If someone doesn't understand something here, I hope it's me. You just cannot keep giving away bits and pieces of the United States and justify it by mouthing inane platitudes about free trade. We deserve better—and besides, it is anything but free.

The next time you hear of a study group going to Japan, ask them to bring back the answer to one question: Why, if Japanese products are so much better than ours does Japan not let our inferior products compete side by side in their stores with their own products? Of course we already have the answer to that one: It would upset Japan's near monopolistic applecart.

Check out the prices of any Panasonic or Sony products in our stores and in Japan's. If the prices in Japan are *only* 50 percent higher than those for the same products here in the States, their consumers are getting a bargain.

Another question that should be asked is "Why are we the only target for Japan's aggressive marketing tactics?" (I might have said *trade tactics*, but that would have been a misnomer.) Why haven't France, Germany, and Great Britain similarly borne the brunt of one-sided competition? The answer is simple: Their governments have moved to protect their industries by enacting domestic-content legislation. As a Goldwater conservative in good standing, this sort of thing is anathema to me, yet even this is preferable to slow strangulation.

I don't believe that mere survival should be our highest interest. That puts us on the defensive. What our posture should be is to once more assume the initiative that we have given away. George Santayana once said, "Those who cannot remember the past are condemned to repeat it." The lessons of history here are crystal clear. Manufacturing and only manufacturing creates wealth. Payment for services only redistributes existing wealth but does not add to it. For proof of this, one only needs to look to the Far East for indisputable evidence. Japan, Hong Kong, Taiwan, and Singapore are all thriving—and all without appreciable natural resources.

On the other hand, England was the global titan throughout the 18th and 19th centuries. Then they rechanneled their energies into becoming the publishing, banking, and educational capital of the civilized world. It didn't take long for their power to wane.

For the life of me, I cannot see what is wrong with a dollar-for-dollar trade policy. Heck, I'd even settle for a *dime* on the dollar—at least that would be 9 cents more than we are getting now. There are two offshoots from this one-sided "trading" policy that are most troubling. First, inspired by the success of the Japanese, practically all the countries in the Orient have now placed restrictions on foreign imports, especially those from the United States. Second, in an unmitigated act of gall Japan is harshly rebuking us for our huge imbalance of trade.

How many administrations will come and go and how much of the United States will be traded away before something is done to reverse this trend is difficult to forecast. But one day we will have to come to our senses and say enough is enough. We are projecting an image of weakness. It is tantamount to Mike Tyson entering the ring with one arm strapped to his side. How many bouts could he survive, much less win? Similarly, how can we flourish by serving as a dumping ground for the world's products when half the world's ports are closed to us?

I believe it is safe to say that this predicament doesn't bear even a distant relationship to Japan's production "miracles." The dollar and the trade-policy debacles are the reasons we are getting the pants beat off us in the would market. Hopefully, it will not always be that way. The United States is still the greatest country the world has ever known—it just likes to pretend it's not.

# APPENDIX B

## INVENTORY ACCURACY AND THE ANNUAL PHYSICAL ABERRATION

The annual physical inventory is a typically American institution—typical because it was born of a presumed need and hung around long after it should have been supplanted by a better system. Of all business practices, the annual physical inventory is the most inefficient, wasteful, and useless. The intended purpose is to satisfy the company's auditors (internal or external) that the counts so laboriously gathered and computed accurately reflect the company's investment in inventory. In fact, most policies and procedures manuals carefully detail the specific steps to be followed when taking a physical inventory.

What usually results from all this counting is to take incorrect balances and replace them with other incorrect balances. This should not really be surprising. Counting inventory is an onerous and boring task. Many of the individuals employed for this purpose have no direct involvement with the results, so they want to dispense with the job as quickly and as painlessly as possible.

The single greatest disadvantage to the annual physical inventory is the tremendous cost in both time and money. Whereas many of the costs of running a business are based on assumptions and projections, the actual out-of-pocket costs of the physical inventory can be computed with great accuracy. There are direct costs and indirect costs, and both deserve consideration.

Many manufacturers allot three days to the taking of inventory. To restrict the amount of downtime, these days are usually Thursday, Friday, and Saturday. You can calculate the value of lost production by simply taking the average daily production times the two (or three) days down. Regardless of the rationalization, lost production time can never be made up. It is lost forever. It is usually necessary to spend several days in preparation for the inventory. Generally this is handled by Production and Inventory Control and workers in the stockroom. While this does not contribute to the overall cost, it does preclude them from performing their normal duties. Material handlers cannot move production while they are aligning stock.

Unless it involves overtime, preparation time is indirect. Manpower, rental equipment, and paperwork, however, are not. Because factory personnel and hourly paid clerical staff comprise a good portion of the counters and checkers, this sixth day of the week is paid at time-and-a-half. (And because they are being forced to work on Saturday, many will not show up on Monday.)

So the disadvantages of taking a yearly physical inventory are many and costly. Inventory is always a pressure situation, which can only result in still more waste and errors. Furthermore, it allows inventory to be updated only once a year for the majority of parts.

Because of all these drawbacks, the current—and only reasonable—approach is the periodic-cycle inventory. Cycle-counting inventories involves taking counts on specific parts at regular intervals during a particular period of time. In this way everything gets counted at least once a year. Higher-value items get checked and verified as frequently as is necessary. Part numbers can be pulled for cycle counting when their stock is at the minimum level. If certain parts tend to be troublesome, they can be identified and the source of their problem uncovered.

This is not to say that cycle counting doesn't have disadvantages. The most challenging is the establishment of paperwork cutoff dates. Since cycle counting is being carried on during normal production, it requires a disciplined approach. Moreover, as inventory reconciliation is difficult even during the *annual* count, thought and planning must go into the system for it to function effectively.

Another problem involves the handling of obsolete inventory. It will still be on your inventory records but will never be included in the parts listed for counting. Unless some provision is made for their scrap or disposal, they could conceivably lie there forever. One company found that a quantity of obsolete generators had accumulated in close proximity to the employee exit. During the ensuing time it took to finally find a buyer, 35 percent of them disappeared.

Perhaps the most vexing and potentially damaging question is on whom the responsibility for cycle counting rests. The common solution is to assign this responsibility to stockroom personnel to be performed along with their regular duties. But when the plant gets busy and stockroom activity picks up, this less productive effort is discontinued. Once it has been discontinued, the system has failed and it is difficult to revive it. The inevitable result is inventory buildup in some areas, discouraging shortages in others, and incorrect balances on nearly everything.

The best approach is to assign the responsibility for cycle counting to a person or team on a permanent basis. If this is not feasible, at least it should be made their primary responsibility, with other duties performed as time permits, not the other way around.

Of all the records a business generates in its day-to-day operations, probably the most maligned is its inventory stock records. The reason for this is apparent. If you were to ask any 10 production executives how accurate their inventory records were, 9 would reply either "very good" or "excellent." They reason that because product is going out the door, customers are being served, and costs don't seem to be out of line, this must be so. The truth is that 5 out of the 10 actually have no idea how accurate their inventory records are (maybe they don't care because they have more pressing problems), but to answer more truthfully would be embarrassing.

The upshot is that too little effort is put into increasing the accuracy of inventory and too many companies fail to recognize its importance. They tend to rely on the annual physical count or cycle counting to straighten out their records. But why should inventories be so abused. Suppose Engineering called out on a blueprint a dimension of 16 3/4 ±10, maybe; or how long do you think a controller would last if he told his boss "last month, I think we made X dollars of profit, but I'm not sure—it could have been a lot less—or it could have been more." These silly examples, however, accurately reflect the precision that exists in all too many inventory record balances. If accuracy demands a high priority in Accounting or Engineering, common sense dictates the same importance be accorded to inventory balances. The problem is not of such a magnitude to likely cause a company to fail, but it is a guaranteed certainty that it will have a negative impact on productivity. Not enough effort is being expended in tightening up paperwork procedures, which are a bother to most people, to be avoided if at all possible. It may take an executive decision at the highest level to get this working. A well-planned MRP system is doomed to failure without the proper backup of information, and there is no substitute for accurate inventory records the first time around.

# APPENDIX C

## SYSTEM ASSESSMENT AND PERFORMANCE MEASUREMENTS

### Inventory Planning

In support of MPS, available resources must be committed. This is accomplished by a series of steps of which material availability is the starting point. Meeting the demands of the MPS mandates effective use and free flow of information to control this function.

### Conditions

- Inventory is a business asset; money has been spent to acquire it.
- As an asset, inventory is expected to produce benefits to the firm. Return on investment.
- Inputs to the inventory file are transaction driven.
- It answers the two basic questions of inventory control: What have we got and where is it? What do we need and when are we going to get it?

### System Assessment

1. Are there existing files that reflect how much inventory is on hand?
2. Do these files also show how much is on order?
3. How often are these files updated to correct balances?
4. Is cycle counting used as an alternative to the physical inventory?
5. Is an annual physical inventory taken?
6. Is cycle counting a daily routine?

7. Are inventory replenishments planned by statistical or mathematical models?
8. Are safety stocks carried for raw materials and component parts?
9. Do you maintain a finished-goods inventory?
10. Do inaccurate lead times tend to increase inventory levels?
11. Does inventory shrinkage foul up your production schedules?
12. When parts remain following a production run, are they returned to stock and the count included in the next inventory printout?
13. Does management mandate a given level of inventory?

### Performance Measurements

1. Percentage of inventory accuracy (number of counts outside parameters divided by total number of inventory items). The goal is 95 percent.
2. Number of inventory turns per year (total of all inventory purchases during the year divided by the amount on hand at year end).
3. Amount of inventory dollars freed by reducing lead time.

## OPERATIONS

How does the MPS trigger orders? In order to convert a customer order into a form usable by production, it is first necessary to break it down into its component parts so that parts requirements can be calculated. The practical and logical way to do this is through the bill of material, which:

### Conditions

- Is a listing of all direct components and the quantity of each required that go into making up one end item.
- Must reflect how the product is put together.
- Must show the relationship of parts to each other and to the next-higher assembly.

- Must permit accurate cost buildups.

## System Assessment

1. Is there a consistent, defined numbering system for parts identification?
2. Is the responsibility for the creation of the bill of material clearly defined?
3. Is there a bill of material for every end item produced, including all assemblies?
4. Do all users of the bill of material work from the same document, including revisions?
5. Is the engineering change level data maintained in the bill of material file?
6. Is the effective date of latest engineering change maintained in the bill of material file?
7. Are the inventory levels (raw materials, work-in-process, and finished goods) considered when determining the date when an engineering change will become effective?
8. Is there a material review board that meets periodically?

## Performance Measurement

What is the percentage of bill of material accuracy? The goal is 98 percent. How does the MPS trigger orders? The MPS contains the totality of customer requirements for a given period. Each requirement is linked to a customer code number (service and replacement parts may necessitate a slightly different treatment), and the bill of material can be thus identified. The bill of material is exploded level by level, and all components are listed by their gross requirements. These are reduced by what is on hand and what is on order, and the difference becomes the net requirements for each individual part. It is the quantity of each net requirement that must be ordered.

## Conditions

- Includes both purchased and manufactured parts.
- Inventory count accuracy is of critical importance at this stage.

- Benefit is gained by planning requirements in advance of needs.

**System Assessment**

1. Are purchase orders tracked to ensure on-time delivery—not early and not late?
2. Do orders issued to the shop contain due dates that reflect the actual date of need?
3. Are lead times for purchased parts or manufactured parts excessively padded?
4. Does Purchasing maintain a hot list for expediting rush orders?
5. Do you maintain safety stocks for critical parts that are transported over very long distances?
6. When a part cannot be made or bought according to schedule and must be pushed into the future, are all companion parts similarly pushed out?
7. Are orders for purchased or produced parts time-sequenced according to need?
8. Do you have a committee for make-buy decisions?
9. Does Purchasing normally maintain several vendor sources for each purchased part?
10. Is there at least daily communication between the material-planning and purchasing departments?
11. Is there a provision for identifying and disposing of obsolete parts?
12. Do system lead times accurately reflect vendor-quoted lead times?

**Performance Measurements**

1. What is the percentage of purchasing performance (number of orders received late—or early—divided by the total number of orders placed)?
2. What is the percentage of manufacturing performance (number of orders completed beyond due date divided by the total number of orders in shop)?
3. What is the percentage of vendor performance (number of orders received late divided by the number of orders

placed with vendor)?
How are orders controlled on the shop floor?

## Condition

The shop floor is the nerve center of the production process. Whatever is done or not done at this state will impact directly on the company's profitability.

## System Assessment

1. Does the routing information on the shop order accurately reflect the production process?
2. Does it frequently become necessary to expedite orders through the shop?
3. Do shop orders contain current standards for setup/run time?
4. Are manufacturing orders dispatched to the shop on no more than a weekly basis?
5. Does Production Control receive feedback from the shop on an ongoing basis throughout the shift?
6. Do bottleneck operations interrupt the efficient flow of parts?
7. Are capacity considerations (load profiles) recalculated and updated periodically?
8. Is access to the stockroom limited or controlled?
9. Is there a program for keeping rework under control?
10. Are quantities on production orders overrun in order to make better use of setup times?
11. Do "spongy" or inaccurate lead times cause frequent schedule changes?
12. Does it frequently become necessary for manufacturing to split orders?
13. Are employees cross-trained on related equipment within a work center?
14. Are standard rates set for each piece of machinery in production?
15. Does floor supervision work to due dates on shop orders?

16. Are shop orders remaining open long last past their due dates?
17. How extensive is unplanned activity?
18. Are work orders closed short with remaining material unaccounted for?
19. Is the material for the floor accurately reflected in the work-in-process system?
20. Are work orders completed while outstanding component shortages remain open?
21. Do the priorities driving manufacturing reflect the MPS backlog?
22. Is there excess inventory on the floor that cannot be accounted for?
23. Is the system used for identifying work order priorities?

## Performance Measurements

1. With respect to the number of shop orders flowing through the system for any given period, determine:

   Percentage of orders being closed short.

   Percentage of orders ending up as split orders.

   Percentage of orders being overrun equals total percentage system is out of balance.

### Quick-Check Performance Review

| | *Poor* | *Fair* | *Good* | *Excellent* |
|---|---|---|---|---|
| Inventory records | 1 | 2 | 3 | 4 |
| Bill-of-material records | 1 | 2 | 3 | 4 |
| Master forecasts | 1 | 2 | 3 | 4 |
| Master production schedule | 1 | 2 | 3 | 4 |
| Production lead times | 1 | 2 | 3 | 4 |
| Vendor lead times | 1 | 2 | 3 | 4 |
| Shop floor control data | 1 | 2 | 3 | 4 |
| Capacity plan | 1 | 2 | 3 | 4 |

## The Delivery System

How is customer demand linked to the master production schedule? To achieve the ultimate goal of customer satisfaction, the grouping of orders comprising the master production schedule must follow a set of rules that properly plan the priority of each order so that the desired end result is reached:

### Conditions

- The master production schedule must be realistic.
- Available resources must be allocated.
- Actual due dates (customer-need dates) must be valid.
- Validity can only be accomplished by proper priority planning.
- Executing the master production schedule as stated is a major company objective.

Evidence of customer demand flows into the MPS from either or both of two sources: forecasting (estimate of future demand) and order entry (ongoing current demand).

### System Assessment

1. Are the need and responsibility for preparing the forecast defined?
2. Does the forecast planning period cover the longest customer-quoted lead time?
3. Does the forecast measure seasonal changes?
4. Does the forecast provide for tracking trends?
5. Is the MPS backlog sufficient to cover the items with the longest lead time?
6. Does the MPS, through backscheduling the customer order, provide for the longest cumulative lead time?
7. Is the current portion of the MPS makeable?
8. Does the customer order backlog contain a large amount of past-dues?
9. Does Marketing coordinate customer needs with Scheduling?

10. Do quoted lead times and delivery promises reflect manufacturing realities?

**Performance Measurement**

1. Percentage of forecast error (estimated demand divided by the actual demand).
2. Percentage of orders past due (current portion of items not completed on time divided by the number of items on current schedule).
3. Average length of past-due orders in days.
4. Dollar value of past-due orders (dollar value of orders completed late during month divided by monthly sales).

How are delivery commitments made? The final tallying of success or lack thereof for a firm engaged in manufacturing is summed up here:

**Conditions**

- The delivery history is a clear statement of a company's competitiveness.
- The delivery history is a clear statement of a company's viability to continue as a top-value competitor.
- This must be management's final and overriding commitment.

**System Assessment**

1. Are customers required delivery dates committed to by Marketing following Scheduling's approval?
2. Does Marketing have final say relative to the sequential run of orders?
3. In case of a conflict in order due dates, are items squeezed into the schedule by compressing lead times?
4. Are service and replacement part orders scheduled on the MPS?
5. Do "unavoidable" schedule changes frequently cause missed due dates?
6. Do you avoid the end-of-the-month crunch?

7. Do orders for off-the-shelf items frequently get back-ordered?
8. Does your delivery history, including quality, exceed industry norms?
9. Are you able to gauge the level of customer satisfaction?
10. Are branch warehouses involved in your distribution process?
11. Do orders from these locations, including interplant transfers, receive the same attention as orders from customers?

### Performance Measurement

Delivery scorecard:

- What is the percentage of orders shipped complete, including service and replacement parts? The goal is 100 percent.
- What is the percentage of orders shipped on time, including above? The goal is 100 percent.
- What is the dollar value of late and/or incomplete orders versus total monthly shipments? The goal is zero.
- What is the dollar value of items on backorder versus total monthly shipments?
- What is the percentage of monthly shipping schedule shipped each week? The goal is 25 percent.
- How late are the orders that are being shipped late?

## HARDWARE AND SOFTWARE

Many production and inventory control systems will work both manually and by computer, but not to the same degree. The computer's ability to quickly give a detailed look at the current situation and project future conditions as far as necessary is a major benefit. This becomes increasingly important as the need for data requirements gets more voluminous and complex. The ability of the manual system to effectively plan and replan while maintaining record accuracy is prone to error. A cost-benefit analysis will reveal that:

## Conditions

- Microprocessing hardware is relatively inexpensive.
- Microprocessing software packages are generally ready to use without extensive modifications.
- Most software vendors provide low-cost training.
- The benefits, while not quantifiable, are obvious. Fast, timely and accurate data generation makes it a primary tool for planning and reporting.

## System Assessment

1. Are production and inventory transactions handled by a computer system?
2. Does the current production and inventory-transaction software handle all your needs?
3. Does the computer track material movement through the plant?
4. Do you maintain a master file for each inventory part number?
5. Is training a main ingredient of your system development?

## Performance Measurement

To provide information for many user functions, the inventory master file should have stored the following data:

| *It must contain:* | *It may contain:* |
|---|---|
| Part number | Low-level codes |
| Description | Safety stock |
| Standard cost | Shrinkage |
| Source | Factors |
| Planned lead time | Product codes |
| Load profiles | Community codes |
| Lot-sizing rules | |

It may or may not include transaction data depending upon software and user choice.

# INDEX

# OTHER BUSINESS ONE IRWIN TITLES OF INTEREST TO YOU

## Forecasting [illegible] for Operations Management

[illegible]

## Competing in the World Class Manufacturing

[illegible]

## Measuring Up
## Charting Pathways to Manufacturing Excellence

[illegible]

## Planning for Quality, Productivity, and Competitive Position

[illegible]

[illegible]